Controlling Electrohydraulic Systems

FLUID POWER AND CONTROL

A Series of Textbooks and Reference Books

Consulting Editor

Z. J. Lansky
Parker Hannifin Corporation
Cleveland, Ohio

Associate Editor

Frank Yeaple

Design News Magazine
Cahners Publishing Company
Boston, Massachusetts

1. Hydraulic Pumps and Motors: Selection and Application for Hydraulic Power Control Systems, *by Raymond P. Lambeck*

2. Designing Pneumatic Control Circuits: Efficient Techniques for Practical Application, *by Bruce E. McCord*

3. Fluid Power Troubleshooting, *by Anton H. Hehn*

4. Hydraulic Valves and Controls: Selection and Application, *by John J. Pippenger*

5. Fluid Power Design Handbook, *by Frank Yeaple*

6. Industrial Pneumatic Control, *by Z. J. Lansky and Lawrence F. Schrader, Jr.*

7. Controlling Electrohydraulic Systems, *by Wayne Anderson*

8. Noise Control of Hydraulic Machinery, *by Stan Skaistis*

Other Volumes in Preparation

Controlling Electrohydraulic Systems

Wayne Anderson

Sundstrand-Sauer
Minneapolis, Minnesota

CRC Press
Taylor & Francis Group
Boca Raton London New York

CRC Press is an imprint of the
Taylor & Francis Group, an **informa** business

First published 1988 by Marcel Dekker, Inc.

Published 2019 by CRC Press
Taylor & Francis Group
6000 Broken Sound Parkway NW, Suite 300
Boca Raton, FL 33487-2742

© 1988 by Taylor & Francis Group, LLC
CRC Press is an imprint of Taylor & Francis Group, an Informa business

First issued in paperback 2019

No claim to original U.S. Government works

ISBN 13: 978-0-367-45135-6 (pbk)
ISBN 13: 978-0-8247-7825-5 (hbk)

**Visit the Taylor & Francis Web site at
http://www.taylorandfrancis.com**

**and the CRC Press Web site at
http://www.crcpress.com**

LIBRARY OF CONGRESS

Library of Congress Cataloging-in-Publication Data

Anderson, Wayne
    Controlling electrohydraulic systems / Wayne Anderson.
        p.   cm. -- (Fluid power and control ; 7)
    Includes index.
    ISBN 0-8247-7825-1
    1. Hydraulic control. 2. Fluid power technology. I. Title.
II. Series.
TJ843.A54 1988
629.8'042--dc19   87-30563
                                                   CIP

Preface

This book is intended for practicing engineers and students who have studied feedback control theory (typically in engineering curriculums). A review of system control theory (with electrical and hydraulic examples) is included to maintain clarity and to give a sufficient mathematical treatment to model the components and understand the systems. Mechanical, electrical, and control engineers involved in component and/or system integration of electronic and hydraulic components will benefit from this book. Component designers as well as end-item control engineers should also gain insight into their own systems from the style of analysis. The flow of information leads from pump-motor operation and sizing with valves to linear and nonlinear analysis of the components and systems.

I have tried to reflect energy and cost efficiencies for both the designer and the system user. The computer programs combined with the test results allow a practical, coherent insight into parameters that control system stability and response. Digital systems are studied through the s, Z, and W domains, as well as through state-space techniques.

The book discusses the pump's role in systems and its use as a power source to a control loop and as a component in a system. Proportional valves and servovalves are sized to systems. Pilot-valve arrangements for these valves are discussed, with emphasis

on the nozzle flapper pilot, arranged either with a feedback wire
to a second stage or as a stand-alone pressure control servovalve.

Servovalves are analyzed in pressure and flow control configurations. Stability in open- and closed-loop (analog and digital)
applications is evaluated for several systems utilizing the pump
and valve component parameters. Response optimization is shown
by computer iterations and several methods of dynamic compensation.

The Appendixes contain computer programs for analysis of
the system's open- and closed-loop dynamic response. Also included in the Appendixes is background information for the microcontroller systems discussed (including the mechanism for
software interrupt routines for establishing digital closed-loop
control).

The book is written in a style that I would have desired when
I entered the industry. It is intended to provide a good understanding of the basics, complemented by an understandable, working knowledge of the "real world."

I am grateful for the support of the following individuals. Dr.
Ronald K. Anderson of Bemidji State University, Bemidji, Minnesota, assisted me in many aspects of the book, including math
routines, composition continuity, and derivation of physical laws
(encompassing both classical and modern control theory). Dr.
Dan Dolan of Rapid City School of Mines, Rapid City, South Da-
kota, encouraged and guided me through the state-space approach
to nonlinear simulation. John Myers and Fred Pollman of Sundstrand-Sauer, Minneapolis, Minnesota, provided me with the software for both the text and artwork, and they supported me during the development of the book.

Wayne Anderson

Contents

Controlling Electrohydraulic Systems

1

Introduction

1.1 INTRODUCTION

Hydraulic systems, as well as many other systems, can be analyzed with a wide variety of computer software. Why then, with
computers becoming very accessible to engineers and managers,
should an engineer need to be well versed on control system theory? The answer is simple: The person performing the system
analysis must have the insight to handle variations which can and
do occur.

For example, a program may be set up to analyze a hydraulic
system controlled by a valve. It may represent a cost improvement, however, to rearrange the components with the pump

1

directly controlling the system. Often in hydraulic systems, low damping, which results in oscillations, must be controlled. The engineer must know how to tie together the practical design locations (for additional damping) with theoretical control considerations, to increase the damping while maintaining adequate response. The control engineer must be able to realize and facilitate such alterations.

Hydraulic components alone can efficiently and precisely control a wide variety of loading conditions. Additional sophistication requires closed-loop control systems, wherein feedback algorithms drive valving arrangements under the scrutiny of controller schemes. Digital algorithms, resulting from microprocessor incorporation, introduce processing flexibility in the electrohydraulic systems that is not obtainable by analog methods.

The evolution of today's quality electrohydraulic systems has spanned several decades. Machining capabilities, pumping, valving and actuator designs, special hydraulic fluids, filtering techniques, transducer technology, and electrical control have all increased in quality and breadth of utilization. Electrohydraulic systems are used in materials testing (by properly fatiguing and vibrating specimens of all shapes and sizes) to ensure the integrity of material, from plastics to special aircraft metals, components, and the aircraft itself. Other areas which benefit from electrohydraulic systems range from machining to (robotic) production automation. Aircraft and military uses are numerous.

System complexity depends upon a variety of factors ranging from end-item responsiveness to cost sensitivity and energy requirements. However, from simple hydraulic components to complicated digitally controlled electrohydraulic systems, stabilized output is essential. The use of control devices to obtain such stability is the subject of this book. The control system theory and applications will be introduced after a discussion of the basic elements which make up the systems.

An example of a hydraulic system is shown in Figure 1.1. Energy from a source produces a hydraulic pressure in the actuator to control the force on and movement of the mass. The energy originates through the rotation of components (pump and valving), is manifested as flow and pressure, and is delivered as a linear motion through the actuator. There is a transition point wherein electronic means of performing this same task is not possible. This point is a function of the actual value of the mass, with static and dynamic constraints.

Some similar processes can be performed electrically (as attested to by the use of electromagnetic shakers, etc.), although a

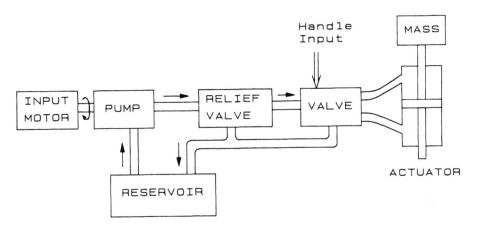

FIGURE 1.1 Hydraulic system.

large mass may be incapable of following the demanded inputs.
Breakdown may even occur because the system is too massive.
Hydraulic shakers, on the other hand, can precisely hold large
masses, with fewer limitations on response (although some hydraulic circuits are much more responsive than others). The actuator could be a rotary actuator or even a hydraulic motor. The
pump can take on a variety of configurations and can be used in
different relative locations, depending on the flow requirements
of the system.

1.2 PUMP, VALVES, AND OUTPUT DRIVES

The pump converts its energy of rotation into a flow usable to
the output drive. In the apparatus of Figure 1.1, the relief
valve sets an upper limit to the pressure in the system. When
the pressure resulting from the pump-motor-load combination
reaches the relief valve setting, oil is dumped to the tank to relieve the pressure. The valving is used to divert, control, or
change the flow into a form usable by the output drive.

The output drive (linear motion actuator, hydraulic actuator,
or hydraulic motor) receives the flow to move the mass in response to the system inputs. The pump also can be coupled directly to the hydraulic motor as a hydrostatic transmission. The

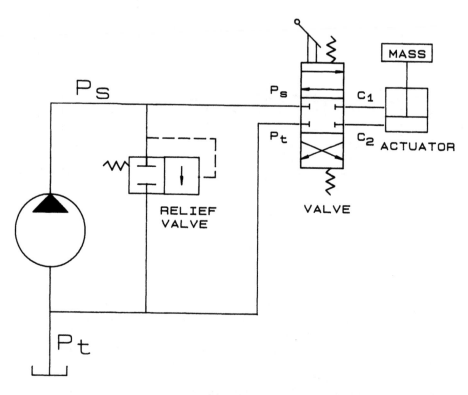

FIGURE 1.2 Hydraulic schematic of Figure 1.1.

valving can use a simple on—off control, or it can employ a complex multistage device with magnetic, electrical, and hydromechanical interfaces. Figure 1.2 is a hydraulic schematic representation for the system of Figure 1.1.

1.3 THE IMPORTANCE OF SIZING

Bigger isn't necessarily better! There are always tradeoffs in
real systems which affect design cost, complexity, and static and
dynamic performance. It would be ideal to be able to have a low-
cost system perform the function of Figure 1.1, with system demands of stroking 10,000 lb by 10 ft in less than 0.001 s. Clearly,

this is a ridiculous situation! Obviously, constraints exist which
do not allow these simultaneous demands.

These restrictions show up, for example, as dynamic lags inherent in the design of a component. Saturation also restricts
the performance of an element. When these constraints of the
components are realized, system requirements demand additional
controls to supplement the overall response.

The valve itself is a controlling element in delivering power to
the actuator. Any component or system is built upon the basic
elements which ultimately define the dynamic lags, saturation
levels, and control limitations.

In principle, high-order systems have more parameters which
could be optimized for improved performance. But high-order
systems are more complex, more expensive, and have more ways
to become unstable. Open- or closed-loop, hydraulic or electrohydraulic systems analysis is based on the physical properties
involved in the components and system. An understanding of the
physical laws governing hydraulic components (and their combinations) is thus essential for deriving and predicting the interactive
behavior and for designing an appropriate control system.

1.4 HYDRAULIC BASICS

Hydraulic systems often make severe demands on the working
fluid. For example, hydraulic components demand oil velocities
of up to 300 in./s. The oil must pass through bearing passages
with clearance of 0.0001 in. and maintain stability throughout a
temperature range from -40°F to over 250°F, often while with-
standing pressures of 3000 psi and higher. Component sensitivity to contamination, especially with tight clearances, demand
good filtration. Contamination and filter sizing are defined and
analyzed in [1].

The main properties of the fluid are the density (ρ), the specific weight (γ), compressibility (K), bulk modulus (β), and viscosity (dynamic μ and kinematic ν). These are defined below for
reference.

Specific weight The specific weight (γ) is the weight of a fluid
per unit of volume.

Density The density ρ is the fluid's mass per unit volume. The
density varies with pressure (P) and temperature (T) approximately as

$$\rho = \rho_0(1 + aP - bT)$$

where ρ_0 is a reference density and a and b are experimentally obtained constraints. Examples throughout the text assume a value of 8×10^{-5} lb-s^2/in.^{4}.

Bulk modulus The bulk modulus of a fluid (β) is the inverse of the compressibility (K), which is defined as

$$\frac{1}{\beta} = K = -V \frac{\partial P}{\partial V}\bigg|_T = -\Delta P \frac{V}{\Delta V}$$

Because of the negative sign, a decrease in volume will be accompanied by an increase in pressure. Typically, the bulk modulus (β) varies from 100,000 to 200,000 psi.

Viscosity The dynamic viscosity (μ) is a measure of the resistance the fluid has to flow. The viscosity decreases with increasing temperature. The kinematic viscosity (ν) is $\nu = \mu/\rho$. Both are extremely dependent on operating conditions and so are generally obtained from tables or manufacturer's specifications. A value of $\mu = 3.6 \times 10^{-6}$ lb-s/in.^{2} is used in this text.

Flow For fluid to flow, there must be a pressure drop. When conditions do not vary with time, the flow is considered "steady" flow. Transients, such as those caused by opening and closing of valves, characterize unsteady flow. The properties which directly affect flow and pressure and the density (ρ) and the viscosity (μ). The nature of the flow can even change as a port is opening, so a progression from orifice flow to even laminar flow is possible. High flow rates may also introduce turbulence. The Reynolds number is a guide to the properties which characterize flow, scaling, and turbulence. The Reynolds number is

$$R = \frac{\rho v D}{\mu}$$

where v is the velocity of oil flow and D is the chamber diameter. In hydraulic systems the oil will flow through pumping, motoring, and valving chambers which vary in size and configuration. The oil is typically ported through hoses, hardline pipes, and holes (within valving and manifolding) with a velocity

$$V = \frac{Q}{A} = \frac{4Q}{\pi D^2}$$

where D is the diameter of the section or hole and Q is the flow rate. For pipes and round hole porting the Reynolds number becomes

$$R = \frac{\rho VD}{\mu} = \frac{4\rho Q}{\pi\mu} = \frac{4Q}{\pi\nu D}$$

The restrictions to flow are typically either viscous forces due to the viscosity of the fluid or inertial forces due to the mass flow rate of the fluid. Experimentally a Reynolds number of 2000 appears to represent a transition point from viscosity-dominated to inertia-dominated flow restriction. For values less than 2000, the flow is considered laminar flow. In this region, systems with the same Reynolds number exhibit scale symmetry.

Laminar flow is characterized as smooth, layered flow restricted by viscosity. This viscosity results in parabolic flow in a pipe. Between $R = 2000$ and $R = 4000$ the flow is in a transition state. Above $R = 4000$, the flow becomes turbulent. For steady flow, the conservation of mass or the "continuity equation" states

$$\frac{dm}{dt} = \frac{d(\rho AV)}{dt} = 0$$

where A is the cross-sectional area through which the oil flows. This means that the mass flow rate (m) is a constant, or

$$\rho AV = m = \text{constant} \quad \text{or} \quad \rho_1 A_1 V_1 = \rho_2 A_2 V_2$$

where subscripts 1 and 2 denote two different chamber sizes. The "momentum theorem" for steady flow relates the external forces on a fluid as

$$F_e = m(V_o - V_i)$$

where

F_e = sum of the external forces

V_o = velocity of the oil out of the flow chamber or boundary

V_i = velocity of the oil into the boundary of the flow chamber

The Bernoulli equation for (nonviscous) fluid motion through a passageway states

$$\rho\frac{V^2}{2} + P + \rho gZ = \text{constant}$$

where Z is the height of the fluid. This can also be written as

$$\frac{V_1{}^2\rho}{2} + P_1 + \rho g Z_1 = \frac{V_2{}^2\rho}{2} + P_2 + \rho g Z_2$$

The term $\rho V^2/2$ is called the dynamic pressure, P is the static pressure, and $\rho g Z$ is the potential pressure.

1.5 THE NECESSITY OF CONTROLS

Hydraulic components, when combined properly to fit their physical interaction and limitations, become a system. Figure 1.3 represents a block diagram of a valve driving a ram with human interface. The portion of the block diagram from the desired handle position to the resulting ram velocity is termed the *open-loop* block diagram of the system. The handle position determines the spool position of the valve, resulting in a flow output of hydraulic oil. This flow rate divided by the area of the ram is the velocity of the ram.

Obviously, for a short-stroke ram, it would not be desirable to operate in such an open-loop mode, because the ram will hit a physical stop unless the input is changed. Even if the input is changed, the open loop allows physical disturbances such as temperature changes or load changes to deviate the flow output from the valve. Drift can also easily occur from valve imperfections, resulting in an output, even though an output may not be demanded.

By visually monitoring the output and altering the system based on the result of observation, one can minimize some of the deficiencies of the open loop. This action is classified as closed-loop control, wherein the input becomes modified for an observed output. The human interface becomes the feedback path of control for the system. The feedback path allows corrective action to occur.

This feedback in the block diagram has actually changed the output from velocity to position (that is, it has integrated the output signal to position). The machine operator initially establishes a handle position (say half of its range) to obtain a final ram position (half of its stroke). Without the visual feedback, the ram will instead reach a velocity representative of the handle (and therefore spool) position and will pass the desired ram position (half of its stroke).

If the operator senses this position visually before or during its occurrence and then cuts back on the handle position toward

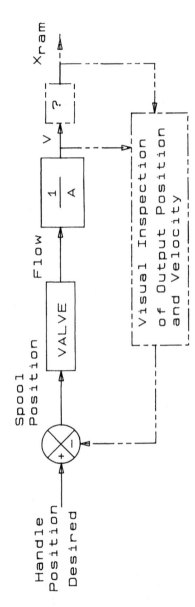

FIGURE 1.3 Block diagram of hydraulic system.

its neutral or zero-flow-producing position, the output velocity will decrease and eventually stop, producing the desired output. The human becomes a means of controlling both the velocity and the position of the output. His or her mastery of commanding and obtaining good closed-loop control (by observation, with thought processes) is limited by the actual response limitations (dynamic lags) of the human body.

The accuracy is actually a function of the visual feedback interpreted by intuition. The effective integration from velocity to position is a function of the response capabilities of the operators in their abilities to change visual observations into a change in handle movement. Therefore the quickness and accuracy of the visual feedback operation will determine the accuracy and response of the system.

For many systems, this visual type of feedback and control mechanism is adequate. However, for more demanding systems with fast-acting servovalves and small-area, short-stroke rams, visual inspection becomes incapable of performing the feedback requirements. If a linkage is provided between the actuator and the spool, a closed-loop interaction allows the handle position to produce a proportional ram position.

The linkage would then become the feedback mechanism. Feedback, in general, allows one to automatically maintain control over the output with minimal efforts. For example, training operators to perform repetitive tasks can be made less fatiguing through automatically closing a control loop.

The feedback link of Figure 1.4 creates a feedback path for the output ram position to be compared with the handle input position; a "closed-loop" system now exists between the valve and ram. For a given input to the handle (say to the right), with the ram position temporarily fixed in position, the spool will be forced to the right by the handle linkage. The spool is ported such that its position reflects oil flow out of C_2 and returns it through C_1. This oil flow produces a higher pressure at C_2 than at C_1. The pressure, when multiplied by the area of the ram, becomes a force pushing the ram to the left.

With the handle position at its desired input, this feedback movement on the handle linkage at the actuator will move the spool back toward its original position. If this action is rapid (small damping), the spool may overshoot the neutral position and reverse the oil flow out of C_1 and return it through C_2; this will create a pressure inbalance in the ram, forcing the ram to the right and bringing the spool back toward its neutral position. This diminishing modulation results in a steady-state ram position

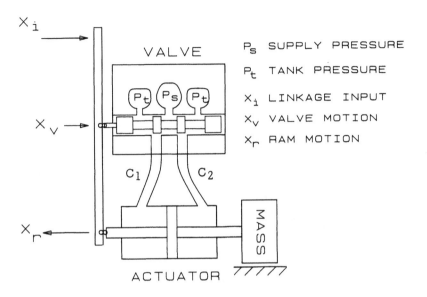

FIGURE 1.4 Linkage feedback for controlling ram position.

which is proportional to handle displacement, although the under-damping may have caused some oscillation in the process.

As the systems become more complex, requirements for additional controls emerge. Figure 1.5 is a closed-loop speed control in which the pump is an integral part of the loop. The previous example used the pump as a source for the pressure and flow requirements of the loop. It operated at a fixed volumetric displacement and dumped any flow not required by the load across the relief valve.

The pump in Figure 1.5 is a variable-volume type with a hydraulic displacement control to maintain its desired output flow. The hydraulic displacement control provides feedback to maintain an output flow which is proportional to the pressure signal input. A change in hydraulic input pressure changes the internal displacement of the pump to produce more or less flow to fit the demands of its closed-loop system.

This closed-loop means of changing flow introduces a dynamic lag not present in the hydraulic system of Figure 1.1. This lag may slow down the system to a point where other means of control are necessary to maintain a responsive system while maintaining

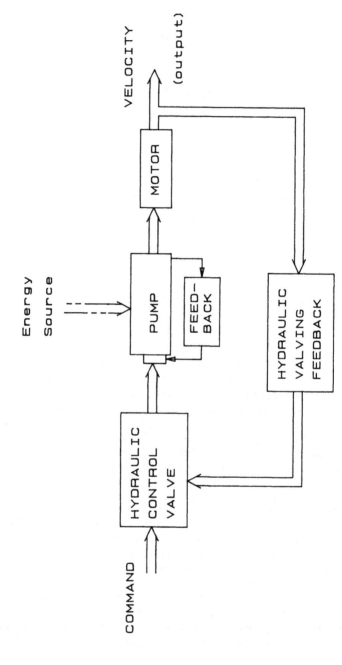

FIGURE 1.5 Hydraulic closed-loop speed control utilizing the pump directly in the loop.

stability. Stability is a primary objective of any system. Dominant dynamic lags in closed-loop systems make the stability requirements even more difficult when response requirements become important. If the combined effects of output velocity and motor output torque requirements are large, the load on the pump and, therefore, its input power source, may become excessive, and will draw down the input speed of the pump.

For a given operating region of the energy source, these loading effects could draw down the source (such as an internal combustion engine) to a stall condition. That is, since the pump is a variable-volume type, it could be "destroked" to produce less flow and therefore less energy to the motor, causing the motor to stop. However, the total system is in a closed-loop mode, wherein the velocity is commanded to a value by the input. In order to keep the energy source within its operating range and to have the ability to destroke the pump under the severe loading, the feedback itself must be able to sense this condition and size the loop accordingly. This type of control action is highly susceptible to system instability.

Hydraulic valving, especially when used for stability compensation, can become expensive. Electrical compensation is also expensive if the electronic elements are not presently employed in the loop. If, however, the electronics are already part of the loop (such as the summing junction of the input command with the feedback, especially if it contains the feedback components), then electrical compensation becomes attractive. Compensation techniques, whether electrical, mechanical, or hydraulic, use physical laws of each of the elements to produce a more stable and responsive system.

Compensation (whether electrical, mechanical, or hydraulic) at the summing junction or in feedback will not perform any better than allowed by the quality of the feedback element. The feedback control element sets the overall scale factor of a system and is largely responsible for the overall accuracy. The scale factor is the actual value range of the output divided by the value range of the input. Therefore, if the output velocity of Figure 1.5 ranges from 0 to 300 rpm and the input ranges from 0 to 6 volts (V), the scale factor is the output divided by the input (300 rpm/6 V, or 50 rpm/V). The inverse of the scale factor is the feedback gain (1 V/50 rpm, or 0.02 V/rpm).

If the feedback gain is inaccurate, the static gain or scale factor will also deviate, since the feedback scales the system. The feedback is demanding because it "forces" the output to match its scaling. The linkage of the valve-ram combination of Figure 1.4, which performed the feedback function, scaled the output stroke of the ram in relation to the input stroke of the spool by the lever ratios of the link.

Inherent component imperfections within the system which distort the feedback can result in poor system dynamic response, especially if uncompensated or optimized with inferior components. Saturation limits of valving and electrical circuitry and dynamic lags are examples of such imperfections. Accordingly, the system only approaches the characteristics of the feedback. If the feedback is erratic or noisy, the total system response will become worse as it amplifies the effect.

Whether the system is load interacting, pump or valve driven, or electrohydraulic actuated, it will be controlled within the bounds of the controller's algorithm (when properly matched by the feedback process). Electrical components are typically used to combine the electrohydraulic components into complete systems. Analog controllers are then typically used in these composite electrohydraulic systems, in closed-loop fashion, to obtain the desired output motion.

1.6 SYSTEM ENHANCEMENT WITH THE MICROPROCESSOR

The microprocessor has the ability to change the control scheme to fit changes within the operating system. In addition to providing information (of various system parameters) on demand, the microprocessor can change a valve drive signal to overcome a random loading situation. The microprocessor must use a continuous analog signal from the process or system being monitored and convert it into discrete (digital) levels (which can be manipulated within the microprocessor).

After this signal is combined with the main input command, the microprocessor performs an algorithm to control the process and to produce the desired output. The physical link between the microprocessor and the process to be controlled must change the signal from digital form back to a continuous (analog) signal.

The controlling algorithm varies from simple proportional control to digital equivalents of analog compensation networks and adaptive control schemes. An example might be to perform an FFT (fast Fourier transform) from the feedback data, and to search the spectrum for response anomalies before providing corrective action. When established limits are approached, failures begin to occur. Controller changes may then become desirable, or variable parameters may need investigation. The microprocessor's flexibility allows it to conform to these requirements. Before investigating the marriage of hydraulics with electronics, we examine the potential of hydraulics.

BIBLIOGRAPHY

1. Fitch, E. C. *An Encyclopedia of Fluid Contamination Control for Hydraulic Systems*, Hemisphere Publishing, 1979.

2
The Power of Pressure and Flow

2.1 INTRODUCTION

Pumps provide the source of energy to be controlled. They are
driven by external means, such as electrical motors, or internal
combustion engines. Their input speeds vary, depending on the
application. Although pumps are the primary source for developing pressure, they do not generate pressure without other circuitry. They provide flow to other components in the system.

If a pump's output flow is open to the atmosphere, the result
is flow only. If the output is connected to a restriction such as
a relief valve, the flow will become compressed in the chamber
(hose); this raises the output pressure. This pressure will increase until the relief-valve setting (or damage) is reached.

An accumulator performs the same function. It will discharge
oil until it has depleted its volume. It also needs a resistance to
its flow to produce a pressure. The accumulator is very useful
for a variety of intermittent applications, but it is not sufficient
for continuous-flow conditions.

The most common type of pumps used for hydraulic systems
are the gear, vane, piston, and screw pumps. The output flow
capacity of any style of pump depends on the volumetric size of
its pumping chamber and its input rotational speed.

2.2 PUMPING MECHANISMS

Pumps are matched to hydraulic motors, linear actuators, and
valves. The type of pump used depends on the system demands,
including circuit usage and loading requirements, working and
maximum pressure levels, fixed- or variable-flow requirements,
pump efficiency, leakage, cost, noise levels, and contamination
sensitivity. Very often different pump styles are used congru-
ently to meet total system requirements.

The gear pump in Figure 2.1 is cost efficient and is typically
used at moderate speeds and pressures. The fluid is directed
around the outer section of the gears, and the mesh between the
gears acts as a seal to restrict flow back to the inlet. Some gear
pumps have pressure-balanced sealing on the gear side face,
which increases the leak resistance. The gear pump is commonly
used for powering auxiliary components or for low-flow and low-
pressure valving.

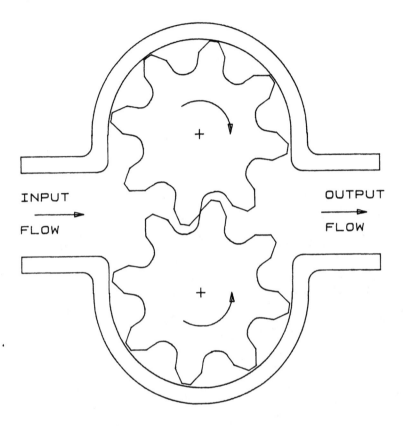

FIGURE 2.1 Gear pump.

The vane pump of Figure 2.2 is relatively inexpensive, tolerant to some wear, and can be used to produce variable flow. The
peripheral leakage is reduced and efficiency is improved over the
gear pump by the outward forces on the vane against the housing.
Variable flow is made possible by varying the eccentricity of the
vane rotor relative to the housing.

Piston-type pumps are manufactured in both radial and axial
designs. The radial piston pump in Figure 2.3 has an inner shaft,
or pintle, which is fixed in position. The pintle has inlet and outlet passageways. The rotor (just outside the shaft) rotates about

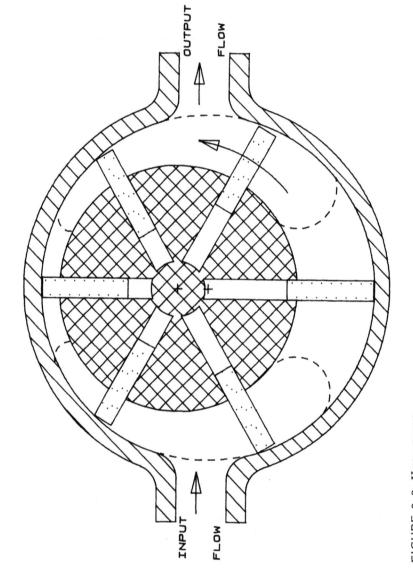

FIGURE 2.2 Vane pump.

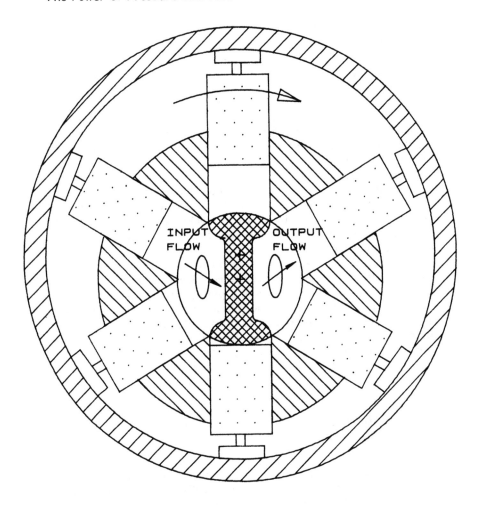

FIGURE 2.3 Radial piston pump.

the pintle, causing the pistons to move radially (because of the eccentricity of the pintle relative to the housing).

The pintle is shaped to separate inlet and outlet ports for proper timing of the rotor movement. As the rotor moves clockwise (starting from the lower piston), the piston goes outward drawing oil into the piston chambers (until the piston reaches a position on the output section of the pintle). As the rotation continues, the oil is forced out the output port. The volume of oil depends on the piston area and piston stroke; therefore, for a given rotor input speed, the flow rate can be altered by changing the eccentricity between the pintle and housing.

The axial piston pump is shown in Figure 2.4. The pistons are also in a rotor but are positioned axially. The input shaft is connected directly to the rotor. In order to obtain varying strokes like the radial piston pump, the axial piston pump uses a swash plate positioned at an angle from the axis of the piston surfaces. The swash plate does not rotate, but it can be changed in position relative to its pivot. As the rotor (and therefore pistons) moves counterclockwise (as shown in the piston cross section), the bottom piston draws in oil as it strokes outward. The end plate (similar to the pintle) separates the inlet and outlet flow by the kidney-shaped ports.

When the piston reaches the top position and starts toward the bottom position (in the same counterclockwise rotor direction), the pistons go back toward the end plate, forcing the oil out the output port. As with the radial piston pump, one can vary the piston stroke, and therefore the output flow, by changing the angle of the swash plate. Keep in mind that even at this point of discharge, supply pressure has not been set up; the pump has only provided a means of driving oil from a reservoir.

2.3 CLOSED CIRCUITS WITH HYDROSTATIC TRANSMISSIONS

The swash-plate positioning for the axial variable-displacement pump is created mechanically, hydraulically, or electrohydraulically. The pressure and flow requirements for this auxiliary function are less than the possible loading effects which a hydraulic actuator, motor, or linear actuator will demand on the main pumping source. In order to obtain a feel for the difference between auxiliary controls and main control hydraulic elements, hydraulic circuitry will be discussed next.

The hydraulic circuitry of a pumping system falls under two basic categories: *open circuits* and *closed circuits*. An interrelated function of the open circuit is the variance between open- and closed-center circuits (these will be examined also). The

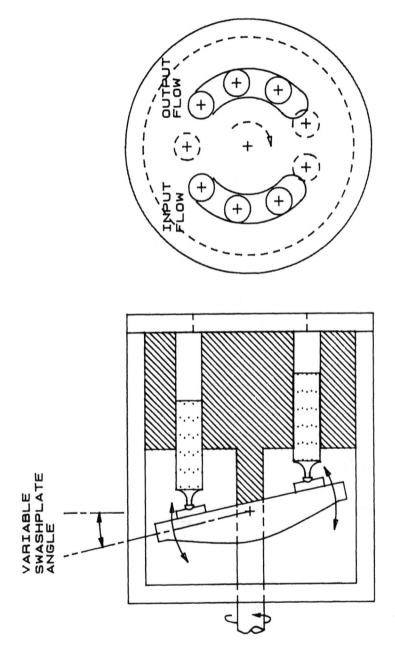

FIGURE 2. 4 Axial piston pump.

main differences between the open circuit and the closed circuit are the interaction between the pump and its output drive and the means of treating the return flow of oil.

The hydrostatic transmission is a closed-circuit system in which the pump output flow is sent directly to the hydraulic motor and then returned in a continuous motion back to the pump. The reservoir and heat exchanger take up leakage and auxiliary function flow. In the open-circuit system, the pump provides flow to other devices, including motors, actuators, and valving. The return flow from these devices is directed to the pump after the reservoir and heat exchanger.

The open-center system differs from the closed-center system in the manner in which pressure is built up for the loading requirements of the system. First we will look at the closed circuit of the hydrostatic transmission. The hydraulic motor, arranged directly as shown in Figure 2.5, is in a typical hydrostatic transmission (a closed circuit). The hydraulic motor works in reverse of the pump. The hydraulic motor has typically the same configuration as the pump, with some modifications.

Irrespective of the type of motor, its function is to transform the flow of oil into torque output to drive a load. Auxiliary hydraulic or electrohydraulic components may be a part of the

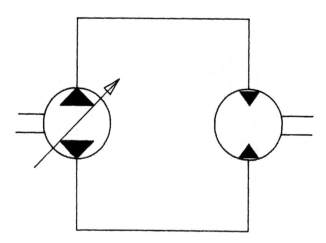

FIGURE 2.5 Hydrostatic transmission.

system in controlling the input, pressure limiting, and other parameters. The motor can also be variable-displacement type. Most of the sections of this book will refer to the axial piston-style pump and motor, but much of the information is applicable to other styles of pumps and motors as well.

The axial piston motor is usually fixed at a particular swashplate angle. When variable, the motor can obtain higher speeds, and sometimes the size of the pump can then be reduced. The variable motor will be described in more detail in Chapter 6. The pump and motor can be coupled together in the same unit; in other cases they may be many feet away.

Leakage plays an important role in pump and motor operation. The efficiency of energy conversion and dynamic response is related to the leakage. Pump-motor sizing is a function of transmitted horsepower and corner horsepower. The transmitted horsepower is that transmitted from the engine to the load (after transients have subsided) for a given load. The corner horsepower is a power limit indicative of the maximum required output of the hydrostatic transmission. It is defined as the product of the maximum torque (required by the motor) and maximum motor speed.

Usually, both cannot be obtained simultaneously; however, Figure 2.6 shows the importance of the corner horsepower by showing both potentials of torque and speed with a single number. The load on a motor is a torque which causes resistance to the pump flow (as it builds up supply pressure on the inlet side of the motor). Thus the motor can deliver either torque or speed as its primary output, depending on the demands of the system. In other words, the system may require a tight grip on maintaining a closed-loop control around torque while speed will vary as needed.

Typically the hydrostatic transmission (pump-motor closed-circuit system) is used to maintain a given output speed proportional to a pump input command. This could be done in the open-loop mode, where the output speed is not monitored other than as speed indication. In the open-loop mode, the temperature effects, engine speed changes, and efficiency changes will force the output speed to vary.

The variable pump must put out a flow proportional to its input. Obviously, for a fixed pump (swash plate fixed at a given angle), the main factor which can change its output flow is the engine speed (or other input energy source). For a variable pump, the displacement (cubic inches per revolution) also can

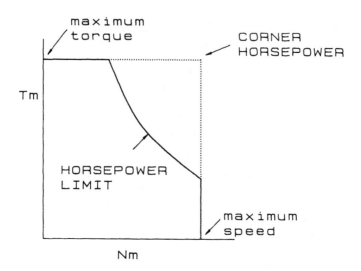

FIGURE 2.6 Corner horsepower indicating maximum torque and speed in the hydrostatic transmission.

be altered to change the flow output. Thus, for a given swash-plate position, the output flow is the pump input speed multiplied by the pump displacement:

$$Q_p = N_p D_p \eta$$

where

Q_p = pump flow

N_p = pump input speed

D_p = displacement of the pump at the swash-plate angle

η = efficiency of the pump

The gain of the pump for a given swash-plate angle θ is

$$K_q = \frac{N_p D_p}{\theta_{max}} = \frac{Q_{output}}{\theta_{input}}$$

where K_q is the flow gain (cis/degree) and θ_{max} is typically 18°.

The swash plate must have its own feedback mechanism to produce proportionality between swash-plate position and the input to the pump. The control for maintaining the position control of the swash plate is obtained by an auxiliary hydraulic circuit; it will be discussed in Chapter 6 after servovalves are introduced. This auxiliary circuit and other valving schemes are achieved by an open-circuit hydraulic arrangement.

2.4 OPEN CIRCUITS AND THE ORIFICE FOR CONTROL

The closed circuit is a continuous circuit between pump and motor. The open circuit uses a pump to produce flow. This flow is utilized by a simple or complex system, with the return flow exhausting to the reservoir. The inlet section of the pump then draws the fluid from the reservoir to complete the circuit. The open circuit produces flow to be utilized in two basic forms: closed-center and open-center valving. The closed-center-valving open circuit will be studied first.

Consider the system in Figure 2.7. The pump could be either variable or fixed displacement. Assuming a fixed-displacement pump with a fixed input speed, the output flow would be constant.

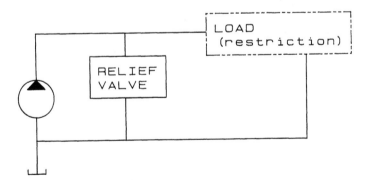

FIGURE 2.7 Relief valve interface with pump and load.

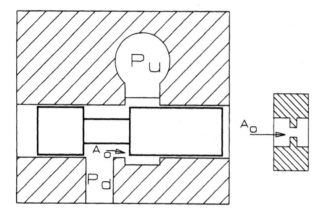

FIGURE 2.8 Variable and fixed orifices.

With an open line to the atmosphere, there would be no pressure buildup. If the output line is capped, the flow will become compressed and the pressure will rise until the hose (or weakest link) breaks.

If a relief valve is attached, the pressure will build up to its setting and dump over the relief and back to tank. Obviously this wastes power and fulfills no useful purpose. If, however, the line is connected to an orifice with a hydraulic actuator, some useful work can be accomplished and controlled.

The orifice equation is applicable in a wide variety of configurations. The orifice is used in valving situations to control pressure and flow, to maintain stability through proper machining between components, and to obtain damping in systems. Figure 2.8 shows two typical orifice styles. In either case, the orifice equation is

$$Q = C_d A_0 \sqrt{\frac{2\Delta P}{\rho}}$$

where

C_d = discharge coefficient of the orifice: if the edge varies from a sharp edge, the flow will vary with temperature

A_0 = area of orifice (inches squared)

    = $\pi D X \beta$ (for the variable orifice)

    = $\pi d^2/4$ (for the fixed orifice)

d   = orifice diameter

X   = spool stroke

β = portion of spool's periphery used for porting

ρ = density of the oil

ΔP = differential pressure between upstream (P_u) and down-
       stream (P_d) pressures

Even at a zero-lap condition, there is flow (either laminar leakage or orifice flow) due to the radial clarance between the spool and bore. There is a transition point within the small annular path in which the flow changes from orifice to laminar leakage flow. The orifice and leakage flows are described by the following equations:

ORIFICE FLOW

$$Q_0 = C_{or} X \sqrt{P_u - P_d}$$

LAMINAR LEAKAGE FLOW

$$Q_1 = \frac{C_{leak}(P_u - P_d)}{L}$$

where

P_u = upstream pressure

P_d = downstream pressure

L   = length of leakage path

l   = leakage

$$C_{or} = C_d \pi D \beta \sqrt{\frac{2}{\rho}}$$

$$C_{leak} = \frac{\pi D B^3 \{1 + 1.5(e/B)^2\}}{12\mu}$$

e   = eccentricity of spool in bore

μ = dynamic viscosity

D   = spool diameter

B   = radial clearance

Notice that these flows are equal at their transition point, so

$$Q_1 = Q_0$$

$$C_{or} X \sqrt{P_u - P_d} = \frac{C_{leak}(P_u - P_d)}{L}$$

Within this overlap region the orifice stroke X is actually the radial clearance B. Therefore solving for L, the equation reduces to

$$L = \frac{C_{leak}(P_u - P_d)}{C_{or} B \sqrt{P_u - P_d}} = \frac{C_{leak}\sqrt{P_u - P_d}}{C_{or} B} = L_t$$

This lap L_t becomes indicative of the transition lap in which the laminar leakage flow becomes orifice flow. At a given pressure drop $P_u - P_d$, any value of annular length (with respect to the metering edges of the body and spool) smaller than the value L_t will reflect orifice flow, where a larger length would indicate leakage flow. This is important in flow simulation and determination of radial clearances. It is not exact, due to valve chamber and spool or body metering effects, but it does reflect adequate performance for valve design.

The relief valve referred to previously used an orifice. The schematic in Figure 2.7 is an open-circuit system which uses a relief valve to set an upper limit to the pressure in the system. The plant could be an effective blocking circuit caused by trying to move an immovable object by hydraulic means or by a hydromechanical failure.

Figure 2.9 indicates the operation of the relief valve in establishing the working supply pressure to the load. The supply pressure, being present at the midsection of the spool, follows passageways to the tank and to the left end of the spool. Flow to the tank does not occur until the spool allows an opening orifice between supply and tank. Statically, this supply pressure is maintained at the left end chamber.

Dynamically the end-chamber pressure will vary because of the fixed orifice present in the path through the spool. With the

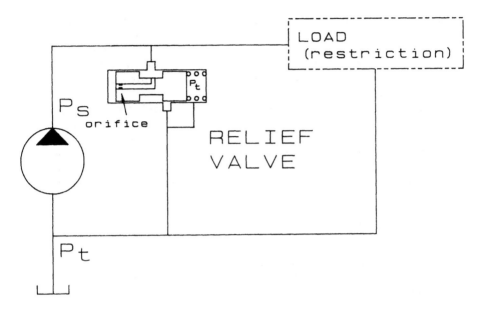

FIGURE 2.9 Relief valve operation with respect to pump and load.

spring biased to a closed-off position between supply and tank,
the supply pressure builds up (for a heavy load) until its static
force at the left end chamber overcomes the equivalent spring
force, resulting in spool movement to the right. The spool movement creates an orifice through which the supply has access to
tank pressure. This causes a lower supply pressure, because
the open orifice has decreased the resistance to flow.

This lower pressure, sensed at the left end of the spool, will
cause the spring to force the spool back toward the left and shut
off the metering orifice between supply and tank. The fixed orifice adds stability to quick-changing, load-induced supply pressures, and the variable orifice performs the actual metering from
supply to tank. Chapters 3 and 6 discuss the actual dynamics of
the valve.

2.5 CLOSED CENTER

Assume that, in addition to the relief valve, a functional system
is used to take advantage of the available power from the pump in

the form of closed-center valves. The valve connected to the
pump and relief valve in Figure 2.10 can build up the supply
pressure from the pump and control this power and thereby control also the movement of the ram (actuator). The valve has four
orifices which can be used with spool stroke to provide flow of
oil to the ram.

In its neutral position, the orifices are "cut off" from porting
oil to the ram. There is, however, leakage flow within the dia-
metral clearance of the spool and its bore. For a given spool position, say to the right, two orifices come into play. The supply
pressure from the pump flows to the right end of the ram while
tank pressure is opened to the left end. Therefore, from the orifice equation

$$Q_1 = kA_0 \sqrt{P_s - P_1}$$

where P_s is supply pressure and P_1 is load differential pressure.

The compressed flow (required to produce pressure) created
by the action of the ram is

$$Q_{1c} \, dt = dV$$

which reflects the volumetric change of the compressed oil. The
bulk modulus of a hydraulic fluid was defined in Chapter 1 as

$$\beta = \frac{V_1}{dV_1} \, dP_1 \qquad \text{or} \qquad dV_1 = \frac{V_1}{\beta} \, dP_1$$

Therefore the change of compressible volume with respect to
time becomes

$$\frac{dV_1}{dt} = \frac{V_1}{\beta} \frac{dP_1}{dt} = Q_{1c}$$

The flow Q_{1c} is that formed by the ram movement against the input flow from the valve. Rearranging to solve for the integrated
pressure, one obtains

$$P_1 = \int \frac{Q_{1c}\beta}{V_1} \, dt = \frac{\beta}{V_1} \int Q_{1c} \, dt = \frac{\beta}{V_1} \int (Q_1 - \dot{X}_r A_r) \, dt$$

Similarly, from conservation of flow, the compressed flow out of
the ram is

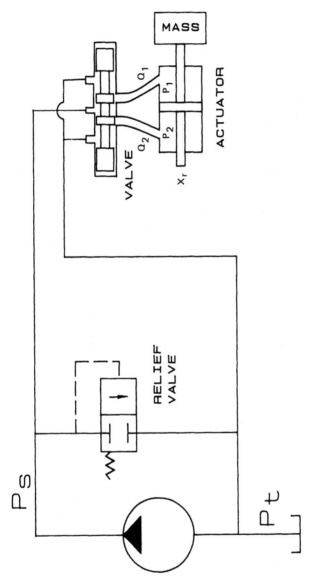

FIGURE 2.10 Valve-actuator load for the hydraulic system of Figure 2.7.

$$P_2 = \frac{\beta}{V_2} \int (Q_2 + \dot{X}_r A_r) \, dt$$

The differential pressure $P_1 - P_2$ is a function of the orifice flow, the bulk modulus of the fluid (β), and the volume (V) of fluid between the valve and actuator. If the velocity of the ram is zero (ram stationary), the pressure P_1 will build up through time to equal the supply P_s. Once this pressure equalizes to P_s, the flow goes to zero, satisfying the orifice equation.

Often a valve is rated at a differential pressure drop across the valve equal to 1000 psi. This has been the convention because the maximum power transfer from a valve to an actuator is accomplished when the pressure drop across the valve is one-third of the supply pressure (typically 3000 psi). The valve has two identical orifices in series; this gives a gain change of $1/\sqrt{2}$ for the single-orifice flow. Thus for a hydraulic fluid with a density (ρ) of 0.00008 and a discharge coefficient (C_d) of 0.6,

$$Q = C_d A_0 \sqrt{\frac{2\Delta P}{\rho}} = 24.6 A_0 \sqrt{\Delta P} \text{ gpm} \qquad \text{for a single orifice}$$

$$= 17.4 A_0 \sqrt{\Delta P} \text{ gpm} \qquad \text{for two orifices in series}$$

The general flow equation is

$$Q = k A_0 \sqrt{\Delta P_v}$$

where ΔP_v is the drop across the orifice(s). The rated flow equation is

$$Q_r = k A_0 \sqrt{\Delta P_{vr}}$$

where ΔP_{vr} is rated pressure (to obtain the rated flow) drop across the orifice(s). Dividing the general flow by the rated flow gives

$$\frac{Q}{Q_r} = \sqrt{\frac{\Delta P_v}{\Delta P_{vr}}}$$

which, solved for ΔP_v, becomes

$$\Delta P_v = \frac{\Delta P_{vr} Q^2}{Q_{vr}^2}$$

The power created by a valve is

$$H_{out} = Q \Delta P_1 = Q(P_s - \Delta P_v)$$

$$= Q\left(P_s - \frac{\Delta P_{vr} Q}{Q_r^2}\right) = QP_s - \frac{\Delta P_{vr} Q^3}{Q_r^2}$$

where $\Delta P_1 = P_s - \Delta P_v$ = load differential pressure and H_{out} = power output of valve. The maximum power transferred can be calculated by taking the derivative of the power with respect to flow and equating it to zero:

$$\frac{d(H_{out})}{dQ} = \frac{d[QP_s - \Delta P_{vr} Q^3/Q_r^2]}{dQ} = P_s - \frac{3Q^2 \Delta P_{vr}}{Q_r^2} = 0$$

Solving for the supply pressure gives

$$P_s = 3 \frac{Q^2}{Q_r^2} \Delta P_{vr}$$

Since $Q = Q_r$ at the rated pressure drop, the rated pressure drop becomes

$$\Delta P_{vr} = \frac{P_s}{3}$$

The single orifice and two orifices in series can control high-power systems. Many valves need the responsive valving action of closed-centered valving with high pressure. The penalties for these systems are the power losses at the relief valve and throttling losses through the valve during the valve porting function. The relief valve, set at its maximum setting with full flow from the pump, will lose this total power (equal to $P_{max} Q_{max}$).

The servovalve output, even though used for maximum power transfer, dissipates one-third of the available power as wasted

heat when porting under maximum power transfer to the load.
Porportional valves and directional valves are more typically
sized for low-pressure drops, thereby operating within the higher flow portion of the load-flow curve. The open-center valve
and closed-center valves, with load-sensing or load-compensating
circuitry, strive to obtain a good balance between power losses
and load demands.

2.6 FROM PROPORTIONAL VALVES
TO SERVOVALVES

The closed-center, open-circuit valve exists in the form of proportional and flow-control servovalves. The difference between
the proportional valve and servovalve stems from the static and
dynamic limits. The servovalve is typically a faster responding
valve. The proportional valve is less expensive with fewer performance demands (such as larger values of hysteresis and null
dead bands) with less valve pressure drop (100–300 psi) rating
(for flow) than the servovalve's (1000 psi) flow rating.

Proportional valves are used primarily in open-loop systems
(systems wherein the human performs the feedback function).
Because of the open-loop operating mode and the proportional
valve's less accurate metering dimensions, the proportional valve
typically has a higher percentage of overlap in the metering function (considerable spool stroke to obtain output flow). The open-
center, open-circuit valve is also a proportional valve.

2.7 OPEN-CENTER VALVE

The open-center and closed-center systems are both open-circuit
systems (the pump provides flow to components with return flow
to reservoir, whereas closed circuits provide a continuous circuit
between pump and motor). An open-center valve is shown in Figure 2.11. It is used with a fixed-displacement pump providing a
constant flow for a given input speed. For no input to the spool
(null position), the spool is positioned as shown.

The supply (P_s) has almost unrestricted flow to the tank (P_t).
The pressure drop is small, and therefore the power loss is small.
Once the spool starts to stroke, the center section of the spool

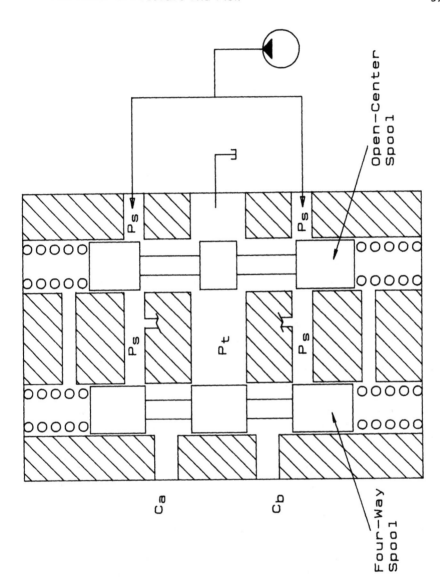

FIGURE 2.11 Two-spool open-center valve separating the four-way and open-center operations.

begins to restrict flow, thereby raising the supply pressure. The spool is symmetric.

For downward spool motion, inlet pressure (top side of open-center spool) is restricted at the inlet edge. The other inlet pressure (of the open-center spool section) becomes larger while the return-pressure (low-pressure) chamber is restricted by the return land of the spool with the body. Further spool movement will close off the open-center spool section and will not allow flow to have a direct route from supply to return.

During this pressure buildup, the four-way spool's supply and return ports begin to meter the pressurized flow to the output control ports (C_a and C_b). With the supply set by the open-center section, the remaining valve function is similar to the closed-center spool discussed in Section 2.4. The metering edges can be shaped and sized to the stroke to provide the proper pressure buildup for the output flow metering.

The main metering edges for output flow (ports C_a and C_b) can be positioned to overlap the function of building up supply pressure in the center section. The plot in Figure 2.12 shows how an open-center valve typically operates for various loads. Two main plots are shown: supply pressure versus spool position, and output flow versus spool position. Approximately the first 20% of spool stroke is developing supply pressure before the actual output flow occurs. The next portion is a combination of both the metering of the open-center spool and the metering of the output flow of the four-way spool.

Obviously, the load changes the profile in the output flow region. That is, the output flow varies with load, which is to be expected. This variation can be minimized by altering the open-center section to blend into actual output flow. Other variations can result in much tighter knit pressure and flow profiles for varying loads. This approaches a goal of pressure compensation, which ideally produces a constant profile, independent of load.

This compensation is created, however, at the expense of wasted power (due to the high pressure buildup during the open-center spool operation). Too much buildup is essentially the closed-center-spool mode of operation, due to high pressure prior to spool stroke (with a larger zero-output zone in the spool stroke).

Figures 2.13, 2.14, and 2.15 represent different arrangements of stacking multiple open-center valves. The valves are designed to be positioned side by side, with porting and spool configurations to allow the functions represented in the flow schematics.

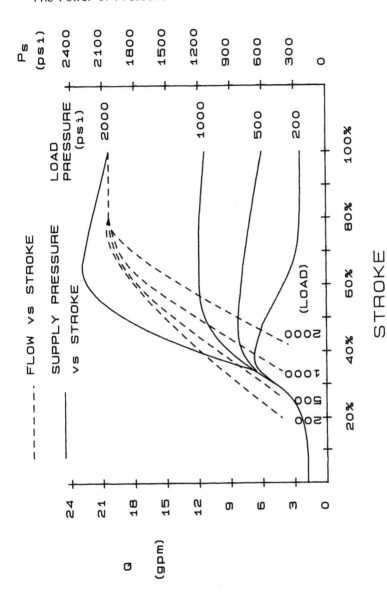

FIGURE 2.12 Open-center valve flow and pressure profile as a function of stroke. The load differential pressure's affect on the flow output is established by the metering matchup between the four-way and open-center functions.

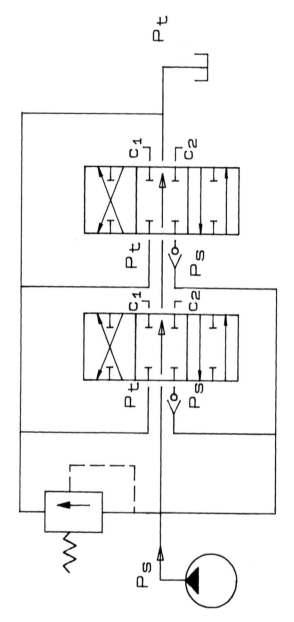

FIGURE 2.13 Parallel open-center spool stackup providing flow to both valves, wherein lower resistance load sets supply pressure when both spools are in their porting stroke.

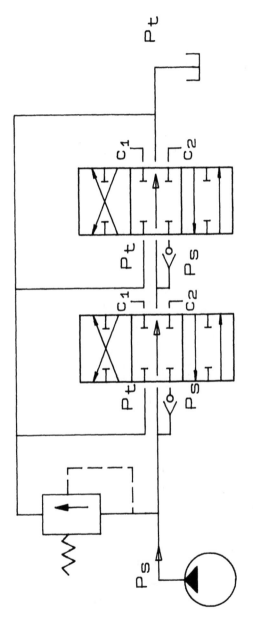

FIGURE 2.14 Tandem open-center spool arrangement. Upstream valve sets priority.

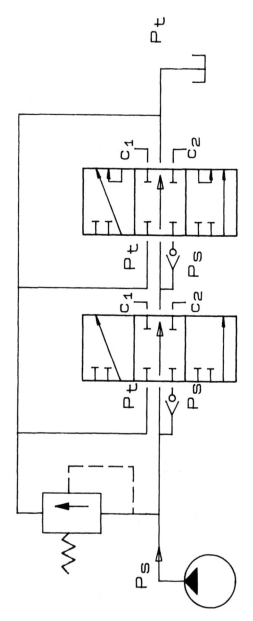

FIGURE 2.15 Series open-center valve orientation. Output of upstream valve becomes source for downstream valve.

The combinations set up priority schemes where multiple functions are encountered. The parallel configuration results in a
supply pressure for both valves set by the spool and its loading
combination. The tandem arrangement sets up a separate supply
pressure for each valve function but is limited by valve section
priorities (the upstream valve has priority).

The series arrangement diverts return flow to the downstream
valve open-center section instead of to tank pressure. Each valve
section will divide the available supply pressure (set at a maximum
by the relief valve), depending on each section's load. With high
demands at each section, the loads can actually demand more than
the pump can provide, causing slower response and less than required pressures at the loads. Sizing the pump to the load, with
the valving configuration, becomes important in fulfilling the system requirements.

2.8 PRESSURE-COMPENSATED VALVE

An effective combination of open-center and closed-center valving
arrangements is the pressure-compensated spool in Figure 2.16.
There are two spools: the closed-center spool, similar to that
discussed in Figure 2.11 (four-way spool), and an unloading
spool, that allows the supply to be dumped or temporarily re-
lieved to tank. The interface between the two spools is the
"logic" built into the main (closed-center) porting spool.

At neutral (no input motion to the spool), the logic is in the
form of a small hole which transfers the tank pressure (load output pressure when the spool is out of the neutral position and in
its metering mode) from the valve body through the spool to the
compensating spool (unloading spool).

This tank pressure is summed with the spring pressure at the
upper end of the compensating spool. This force tends to move
the compensation spool to the base, which closes off flow (from
the supply inlet to return pressure). The supply pressure is
ported to the left end of this spool. With full pump flow and no
metering path to return or output ports (ports C_1 and C_2), this
inlet flow compresses and creates a higher pressure at the inlet
(and therefore on the base end of the spool). This higher supply pressure will eventually match and overcome the spring and
metering pressures (return pressure for the main-spool neutral
position). This will force the spool to the top and unload the
supply to return through the open metering.

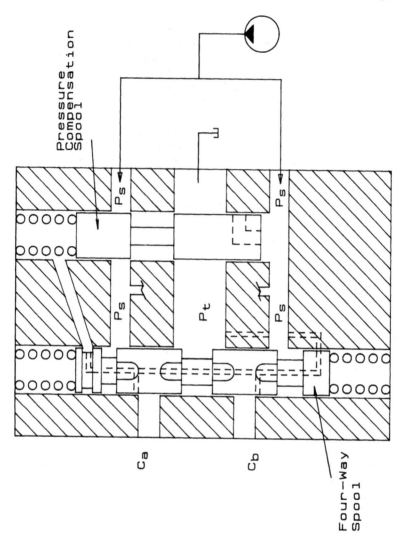

FIGURE 2.16 Pressure-compensated valve. Compensation spool sets supply (P_s) based on load (or tank when four-way spool is at neutral) and spring force.

This spool action modulates back and forth, producing a supply pressure for the total pressure-compensating valve which is equal to the metering pressure (return pressure at the main spool neutral position) plus the pressure equivalent of the spring in the unloading section. When the closed-center spool is stroked out of its neutral position and begins metering flow, the return port is shut off from the compensating section and the metering pressure of the load flow (actuator or motor) is monitored. The result of the compensation is that the pressure is adjusted (by relieving flow) to just match the demands of the load. The spring's pressure equivalent is typically 200 psi; therefore the supply will be set at 200 psi above the pressure required to move the load.

Figures 2.17, 2.18, and 2.19 are different configurations for pressure-compensated spools. The parallel version, by a shuttle valve, chooses the highest demand pressure (metering pressure of each valve section) to establish the supply pressure for all the valves. Therefore only the valve section with the highest output pressure demands will be compensated. The other sections will behave like closed-center valves with high-pressure drops across the metering orifices.

The tandem arrangement stacks up two unloading valves for establishing the supply pressures for each valve section. The unused flow from the upstream pressure-compensator valve section (unloading spool) becomes the source for the downstream compensator. This arrangement, similar to the standard tandem arrangement, sets up a priority scheme for the valve sections that depends on the loads. The series upstream valve ports its return flow, along with the unloading section's unused flow, to the secondary pressure compensator. This divides available supply pressure while compensating each section (as long as the sum of the load demands does not exceed the pump's flow and pressure capabilities).

2.9 PRESSURE-REDUCING VALVES

Many systems, from hydraulic test benches to sophisticated industrial machinery, use pressure-reducing valves. They are used in systems which have multiple functions. If there exists a pressure in a portion of a circuit which is too high for that function, the reducing valve will create a reduced supply.

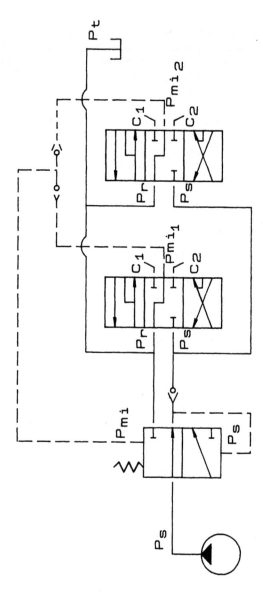

FIGURE 2.17 Parallel pressure-compensated valves. Highest load pressure sets supply pressure for system.

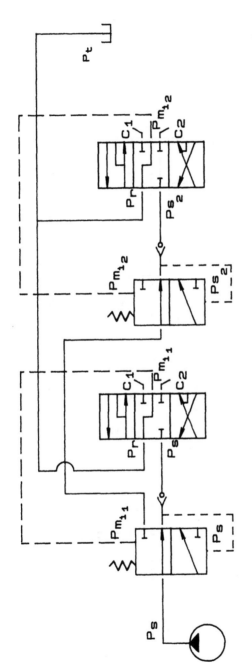

FIGURE 2.18 Tandem pressure-compensated valves. Each section is individually compensated. Priority is set by upstream valve.

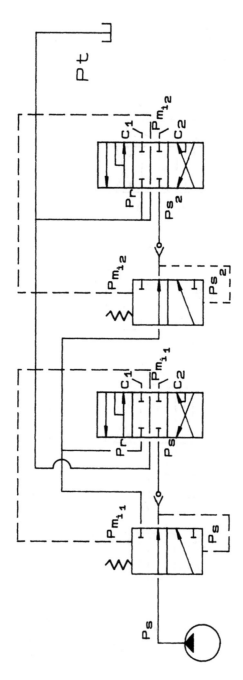

FIGURE 2.19 Series pressure-compensated valves. Output of upstream valve becomes source for downstream compensator and its flow valve.

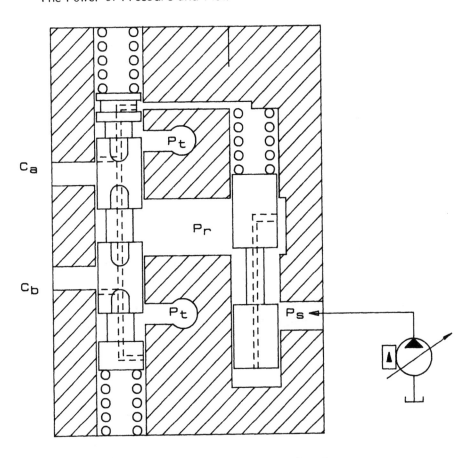

FIGURE 2.20 Pressure-reducing valve. Like the pressure-compensated valve's supply (P_s) of Figure 2.16, reduced supply (P_r) is set by load pressure (or neutral spool-position tank pressure) and the spring bias.

The closed-center spool in Figure 2.20, ported similarly to the pressure-compensated valve of Figure 2.16, relays its required metering load pressure to a pressure compensator. Instead of dumping the supply pressure to return, the supply pressure is metered to a reduced level dependent on the load. The valves of this system must be properly sized to the pump so that the pump can provide sufficient flow for its valving and load functions

without drawing down the supply or reduced supplies to a lower
than acceptable level.

A system may have a single pump driving several functions
which individually desire compensated supply pressures. Shuttle
valves allow the higher of two pressures at a junction to become
the output of the junction. Loads of various valves can be compared to eventually set the main supply pressure, at an unloading valve, to the demands of the highest pressure. The valving
function(s) with lower demands on pressure could use a pressure-
reducing valve.

2.10 LOAD SENSING

The pumps discussed have been typically positive displacement
pumps. The pressure-compensated pump of Figure 2.21 provides
a method of changing the flow at the pump instead of after the
pump. Combined with the pump is a hydraulic servovalve which
monitors the needs of the pump and modifies the pump's swash-
plate position; this changes the oil flow to meet the demands of

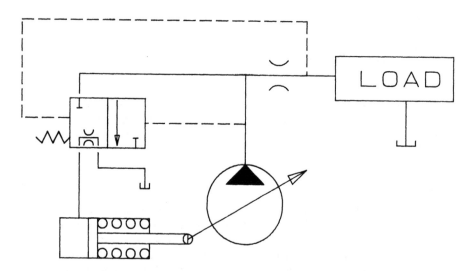

FIGURE 2.21 Pressure-compensated pump. Pump flow established
by pressure drop across orifice (upstream of load).

the load. Downstream from the pump is an orifice. The pressures before and after the orifice are monitored and sent to a pump compensator located at the pump.

This compensator compares the monitored pressures and adjusts the swash-plate position to produce the correct flow output in order to maintain a desired differential pressure across the orifice. With tank pressure downstream from the orifice, the compensator sets a constant flow, which is

$$Q = kA_0 \sqrt{P_s - P_t} = kA_0 \sqrt{P_s}$$

where $P_s = P_{comp}$ (compensator pressure). Therefore the supply pressure is set by the compensator setting. If a load pressure exists downstream from the orifice, the compensator will still maintain the same flow at the same differential pressure. The flow remains the same, but the pressure levels have increased by

$$Q = kA_0 \sqrt{P_s - \Delta P_{load}} \quad \text{where } P_s = P_{comp} + \Delta P_{load}$$

This compensator action occurs if the system produces enough flow to meet the demands of a load. If the load changes, the pump maintains the flow demands of the orifice by raising the supply pressure. If a closed-center spool valve is used instead of the fixed orifice, then Figure 2.22 is the result. The configuration is, for a spool in its metering range, a variable orifice instead of the fixed orifice. For a given spool position and the same compensation setting on the pump control, the resulting flow differs only by the equivalent of the two valving orifices in series. For varying loads and spool strokes, the pump and compensator will provide the necessary flow and supply pressure to meet the demands of the load:

$$Q = kA_0 \sqrt{P_s - \Delta P_{load}}$$

where $P_s = P_{comp} + P_{load}$, which is set by the pump control, and $A_0 = f(x)$ [this orifice is set by spool position (x)]. To better understand this variable-volume pressure-compensated pump control, we will study the block diagram of Figure 2.23. The main objective of the total control system is to maintain a ram position proportional to handle position. This would be the closed-loop goal of the system. The "position transducer" would be a human, monitoring the position and adjusting the handle input control to

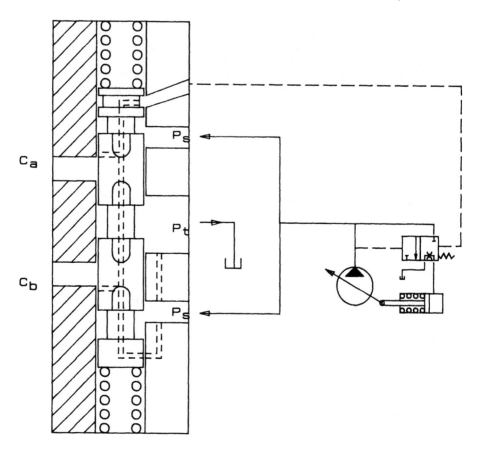

FIGURE 2.22 Pressure-compensated pump with variable orifice
(four-way valve) replacing fixed orifice of Figure 2.21.

maintain the output position. The open-loop output is the velocity of the ram for a given handle position. The role of the pressure-compensated pump is to provide flow proportional to stroke
and independent of load pressure. The flow equation

$$Q = kA_0 \sqrt{\Delta P_v} = kA_0 \sqrt{P_s - \Delta P_{load}} = kA_0 K_{comp} = kA_0$$

states that the flow output is a function of the orifice area and
the differential pressure drop across the orifices. However,

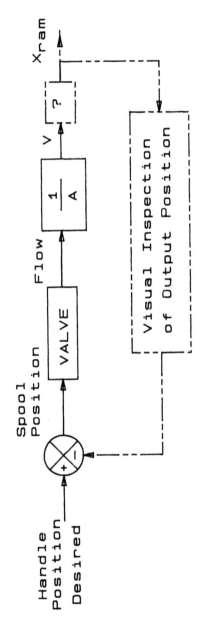

FIGURE 2.23 Ram velocity-position control.

for the pressure-compensated, variable-displacement pump, the
pump control's inner loop provides a constant different pressure
(P_s - ΔP_{load}) across the valve (K_{comp}). The flow, therefore,
reduces to a function of the orifice only. The unspecified block
represents the integration from velocity to position forced by the
monitoring of position. This integrator will be discussed further
in Chapter 3.

Figure 2.24 represents a load-flow profile for the pump valving
characteristics in terms of useful and wasted power. The open-
center valve with no load has minimal power, with loss set by the
the central open-center metering edges. As shown in Figure 2.12,
the design can vary, trading off power loss for better flow output
profile. In the operating ranges for less than full-spool stroke,
the losses can become significant at high loads.

The closed-center valves, if operating at the maximum power
transfer to the load, waste one-third of the supply pressure
(through orificing pressure drops). If the throttling pressure
drop is low with a large load pressure, the losses decrease but
the flow is limited by the envelope of the load-flow curves. Sizing
the pump with the load through the servovalve is discussed in
Section 2.11.

The load-sensing, pump valving combination effectively keeps
a closed-center valve at low throttling losses for any load; this

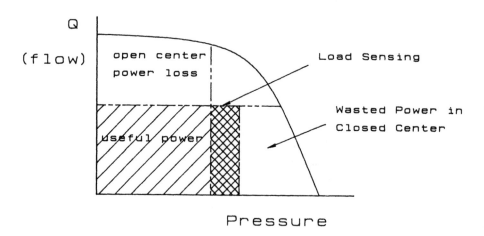

FIGURE 2.24 Pump-valving load-flow profile.

reduces the potential losses. The load sensing enjoys the benefits of both the open- and closed-center valves. In addition to
being powerwise efficient, the load sensing produces proportional
valve output flow (to its input command) because the valve pressure drop remains essentially constant. Since this is a constant
pressure drop, the orifice flow profile becomes dependent on the
spool stroke.

Although the load-sensing circuit has the advantages of matching pumping and loading requirements with minimal power losses,
it is not the most responsive circuit. In other words, the pump,
compensator, and spool do not react as quickly as may be desirable for more demanding systems. Typically this pressure-compensated variable-volume pump is used in open-loop style, although it becomes closed loop when the human interface is sensing
feedback. A human can react to changes only within his or her
physical limits.

This pump configuration is well suited because the human interface and the typical load requirements do not require better
accuracy and speed of response. Even the hydrostatic transmission's closed-circuit systems are not very responsive, but
they are adequate for closed-loop control. A load-sensing electrohydraulic servovalve has been developed [1] which adds servo
response at proportional valve costs. If the pump acts more as a
source of energy for steady inputs, then the system can respond
much faster. The dynamic characteristics of the pump are typically much slower than those of the valving arrangements.

The valve in Figure 2.25 is an electrohydraulic servovalve
driving an actuator. It is not very responsive either, because
of the massive armature. It is represented in block diagram form
in Figure 2.26. The desired output is flow, and the input is an
electrical signal. The position transducer monitors the spool position and converts it to an electrical signal (for comparison with
the input). The error signal drives the electromagnetic circuit
of the servovalve. The result is a movement of the armature.

Directly attached to the armature is the spool. The resulting
flow is proportional to spool position. The spool ports oil to the
actuator (as described previously with the closed-center, open-
circuit valving). The flow divided by the area of the ram is
ram velocity. For a given input voltage (or current) to the
valve, the ram output will travel (at the velocity dictated by
the valve flow) until it reaches its physical stop or until the
input is changed.

The proportionality between the input electrical signal and the
ram velocity is grossly affected by interactive forces on the spool

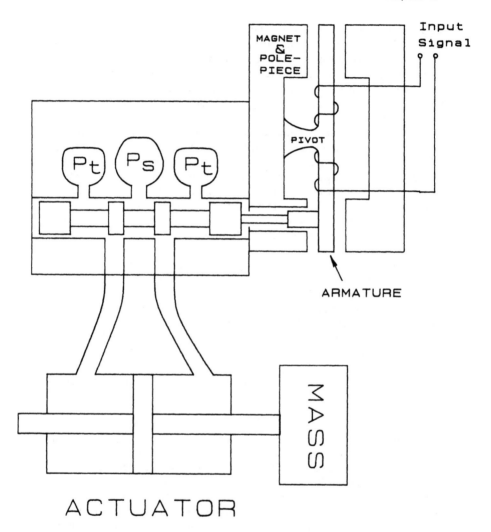

FIGURE 2.25 Flow control.

and load, temperature, and pressure effects between the spool and ram. The inner-loop position transducer will change the output from velocity to position and reduce the external disturbances on the spool and ram.

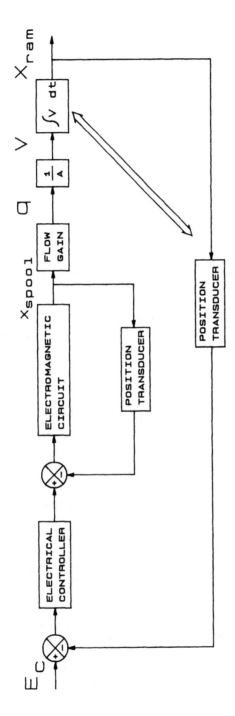

FIGURE 2.26 Closed-loop position control.

The outer-loop position transducer, which acts as an integrator, transmits an electrical signal proportional to position. With
the exception of quickness, it is inferior to the human interface
discussed in Chapter 1 wherein the human had, as a minimum,
control over both velocity and position. The transducer also
sets a scale factor for the combination of the electromagnetic
circuit, spool porting element, and actuator receiving unit. The
position transducer relates x inches to y volts, producing the
scale factor (S.F.). The scale factor is the overall electrohydraulic system static gain (output divided by system input):

$$S.F. = \frac{x}{y}$$

which is the inverse of the feedback gain. This and the dynamic
response are discussed in detail in Chapter 3. Mating the valve
to the ram and load is as important as mating the pump to the
valve. Figure 2.26 shows an inner loop which also has position
feedback. The inner-loop position transducer is less important
than the outer-loop position feedback, because there are fewer
problems due to variations in the inner loop.

2.11 SIZING THE VALVE TO PUMP AND LOAD

Figure 2.27 represents the basic elements of a plant. The pump
input can be an electric motor or internal combustion engine. The
internal combustion engine is discussed in Chapter 6 in a speed
control system. The motor typically must be sized and/or designed to have minimal droop (maintain a stiff link in interfacing
with the load) during normal operating conditions of the pump.
Heavier loading will eventually pull down a motor.

The load, whether acting through a linear actuator, rotary actuator, or motor, is a differential pressure load to the flow-control servovalve. The pressure-control servovalve drives the
same types of actuators and motors but interacts differently with
loads. The pressure-control servovalve and its load dependence
are discussed in Chapter 4. The flow-control valve's load-flow
curve (similar to Figure 2.28) represents the capability of controlling flow provided by the pump.

The envelope of the curve represents the orifice equation and
its square-root relationship to differential pressure (across the

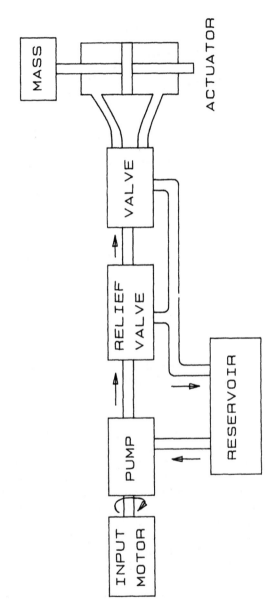

FIGURE 2.27 Hydraulic system.

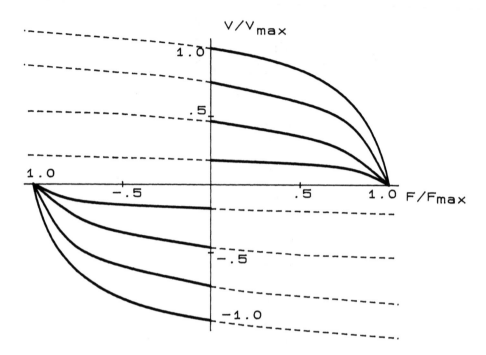

FIGURE 2.28 Flow-control servovalve load-flow curve. Each of the 8 curves represent spool strokes of multiples of ±25%.

valve) and some saturation effects of the passageways (inherent in the valve). The actuator (or motor) is sized to the requirements of its loads as limited by the valve and its capabilities.

The limits are stated statically through plots such as the load-flow curve, and dynamically through time response (frequency response is discussed in Chapter 3). The actuator (or motor) may have requirements demanding simultaneous outputs of force and velocity. The locus of these load demands on a plot of flow versus differential pressure represents the power which must be provided by the pump and valve.

The load-flow curve (of a given servovalve, such as Figure 2.28) becomes the key to matching load requirements to the servovalve. Although the load appears in the form of pressure (force) and flow (velocity) demands, the specific utilization of the valve with respect to the output and its requirements dictates the interface.

If the output is a rotary motor with requirements of maintaining a constant-speed drive, the valve must provide, simultaneously, the required velocity (flow) to the motor with enough
torque (pressure) to match the expected loads that the motor
will see at the maximum speed. If operating cost and fuel efficiency are more important than dynamic performance, this servovalve could be undesirable because it would waste power (due
to the pressure drop across the metering edges).

A better approach would be to use a hydrostatic transmission
with variable-pump delivery. The best match between load demands and valve potential is obtained by enclosing the load requirements within the load-flow envelope, without any void between the valve supply and load needs. With the constant-mo-
tor-speed example, the voids become excessive. Sinusoidal input control over rotary motors or linear actuators, as well as
general position control of linear actuators, can be sized to the
appropriate servovalve. Chapter 3 discusses sinusoidal inputs
for system performance.

The similarities between time and frequency response are both
related to quickness of response and ability to obtain the response
under stability limits. In other words, a ram driving a mass under position control has demands in terms of changing from one
position to another in a certain time period. This dictates both
velocity (flow) and force (pressure) from the valve, similar to a
ram under sinusoidal position control in which a mass is to be vi-
brated at a desired sinusoidal rate. The maximum power transfer between a valve and load was shown to occur when the load
used two-thirds of the supply pressure. Therefore, the desired
point of tangency between the load requirement and the valve
load-flow envelope would be the point of maximum power transfer.

In many cases, the match between the load and valve will not
be tangent at the maximum power transfer. A method of optimizing the load to a valve under sinusoidal positioning has been
developed [2]. The resulting tangency point between the valve
and load requirements is optimized within the "best" operating
range of the valve. The dashed portion of the curves of Figure
2.28 indicate the valve operation when the load effectively becomes the source. In this region, the velocity can increase to a
value larger than the maximum velocity established by the servovalve.

The curves enveloping approximately 80% of spool stroke and
95% of maximum velocity become a target region to avoid the abrupt
change in slope at maximum load (and zero velocity) and to keep

the spool stroke from saturation during transients. The load or horizontal scale varies from 0 to 100% of supply pressure. For a given ram area (A_r), the force developed by this load pressure is $F = F_1 A_r$. Thus, the maximum load pressure occurs at $P_1 = P_s$, resulting in a maximum ram force of $F_{max} = P_s A_r$. Similarly, the vertical (flow) axis varies from zero to Q_{max} (100% of servovalve flow), resulting in a ram velocity ranging from zero to $V_{max} = Q_{max}/A_r$. At maximum load, the velocity is zero, and at maximum velocity the load is zero.

Normalizing the actuator's force and velocity yield the following definitions for F' and V':

$$F' = \frac{F}{F_{max}}, \qquad V' = \frac{V}{V_{max}}$$

where F is a given force requirement of the ram and V is a given velocity requirement of the ram, $0 <\ = F' <\ = 1$, $0 <\ = V' <\ = 1$. In matching the flow and differential pressure load requirements of the valve with respect to sinusoidal inputs to the valve, the actual output flow (Q_a) and load differential pressure (P_a) are needed. They are defined by

$$V = V_s \cos(\phi), \qquad F = F_s \sin(\phi)$$

where V_s is the sinusoidal amplitude of actuator velocity and F_s is the sinusoidal amplitude of actuator force. The normalized valve parameters, with respect to the maximum flow (Q_{max}) and load (P_{max}) become

$$F'_s = \frac{F_s}{F_{max}}, \qquad V'_s = \frac{V_s}{V_{max}}$$

The concurrent sinusoidal requirements of actuator force and velocity can be represented by the vector which has the normalized rectangular coordinates of F'_n and V'_n, where $F'_n = F'/F'_s$ and $V'_n = V'/V'_s$. These normalized values represent the ratios of required actuator force and velocity relative to the maximum force and velocity obtainable. The resulting vector has an amplitude given by

$$\text{amplitude} = (F'_n)^2 + (V'_n)^2 = \left(\frac{F'}{F'_s}\right)^2 + \left(\frac{V'}{V'_s}\right)^2$$

$$= \left(\frac{F/F_{max}}{F_s/F_{max}}\right)^2 + \left(\frac{V/V_{max}}{V_s/V_{max}}\right)^2$$

$$= \left(\frac{F}{F/\sin(\phi)}\right)^2 + \left(\frac{V}{V_s/\cos(\phi)}\right)^2$$

$$= \sin^2(\phi) + \cos^2(\phi) = 1$$

Therefore, since $(F'/F'_s)^2 + (V'/V'_s)^2 = 1$, the load requirements define an elliptical envelope. To stay within the boundaries of a given valve's load-flow curve and to maximize the match between the valve and load, the ellipse should be tangent to the load flow within 80% of maximum velocity and 95% of maximum force. The tangency between the source and load can be determined graphically. Whether the valving combinations are used in sinusoidal, step, or random system usage, the load-flow curve becomes the tool for sizing the load demands to the available power source and valving. Stability and load-source-matched performance are maintained when the sizing is maintained within the envelope discussed.

2.12 CONCLUSION

Through this review of pumps, motors, and valving, it should be apparent that sizing of the components is critical to system operation. The system's responsiveness to input and loading disturbances is keyed to static and dynamic parameters inherent in the components. Obviously, a large variable pump driving a large motor will be slower in reaction time than a small pump-motor system. We must be able to predict component (and therefore system) performance by understanding the parameters and interactions within and between the components. This understanding must follow the physical laws of nature and can be mathematically described. The mathematics, if represented by conventional block diagrams, is very intuitive and fits an unending variety of systems.

BIBLIOGRAPHY

1. Hogan, Brian J. Two spools create three-way, load-sensing servovalve, *Design News*, Vol. 421 (1986), 140–141.

2. Clark, Allen J. Sinusoidal and random motion analysis of mass loaded actuators and valves, MTS Systems Corp., Minneapolis, Minn., 1985.

3

Control Theory Review

3.1 INTRODUCTION

Mathematical model evaluation of system components is necessary
for optimal, stable control. The control engineer must strive for
optimal control of a system within the realm of component complexity, energy levels, and interfacing with existing apparatus. Optimal control is dictated first by stability criteria and is complemented by proper sizing of control parameters. Although these
parameters can be determined by testing, the control engineer
should be able to derive theoretically the mathematical models
and predict system behavior before actual construction.

   This chapter is a summary of control theory applied to electrohydraulic systems. A complete treatment of control theory
analysis following this book's nomenclature can be found in [1].

   The Laplace transform is a very effective way to analyze control systems. It is a mathematical method of relating the output
of a given component and/or system to its input command. We
discuss the convolution integral to show how the Laplace transform can be used to simplify the mathematical treatment of systems. The application of valving in Chapter 2 is expanded to
show its block-diagram model and its Laplace-transform representation.

   Assume that the pump and relief valve in Figure 3.1 treat the
system as an ideal source. It remains to study, predict, and control the dynamic behavior of the valve and actuator. The simplified block diagram of Figure 3.2 shows that a given input X_v to
the valve results in oil flow (output) from the valve into the actuator. Ram velocity (flow input divided by ram differential
area) therefore results from the initial input to the valve.

   We might intuitively raise several questions about such a system. First, it is readily apparent that with no change in input
X_v, the ram would eventually bottom out at one end of its stroke;
therefore, how can the ram speed be properly controlled (especially if the ram stroke is short)? Second, how will the valve and
ram react with loads? Dynamic interaction between the valve,
ram, and load are discussed after the Laplace transform is re-
viewed.

   If the valve and ram were altered (by including the linkage)
to the configuration of Figure 3.3, the closed-loop block diagram
of Figure 3.4 occurs. For a given input motion X_i, to the left in
Figure 3.3, the linkage will pivot about a fixed point at the ram
connection because the ram is presently fixed. This linkage movement will cause the valve spool to move to the right, allowing oil

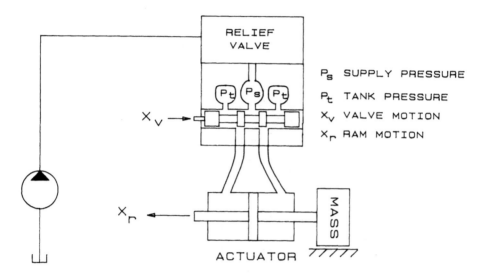

FIGURE 3.1 Hydraulic system.

to flow out the valve to the right side of the actuator; the return
flow from the actuator will flow into the valve, to tank pressure.
    This high-side pressure on the right of the ram will cause the
ram to move to the left. This motion causes the linkage to move
to the left, pivoting about the input portion of the link (no
change in input requirements). This total linkage movement

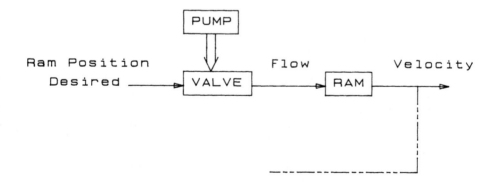

FIGURE 3.2 Valve producing ram velocity.

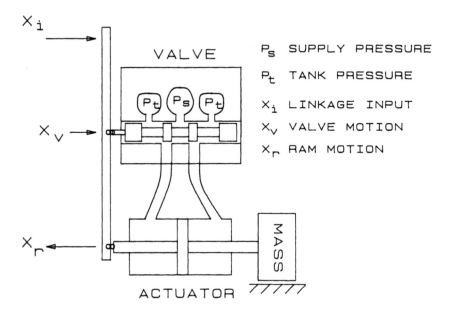

VALVE
ACTUATOR

X_i

X_v

X_r

MASS

P_s SUPPLY PRESSURE

P_t TANK PRESSURE

X_i LINKAGE INPUT

X_v VALVE MOTION

X_r RAM MOTION

FIGURE 3.3 Position control. Linkage feedback integrates velocity of Figure 3.2.

causes the valve spool to move in the direction opposite to that
commanded by the input X_i. This closes off the flow of oil to the
right of the actuator to the point where it will keep the ram stationary. The net result is that the ram position has changed to
a location proportional to the input command X_i. The linkage has
caused the valve-actuator system to become closed loop and has
allowed the ram to obtain a more meaningful "position" proportionality to the input (rather than the velocity proportionality of Figure 3.1).

Figure 3.4 is a general block diagram of the system with X_i as
input and X_r as the resulting output. The linkage "forces" the
output to go from ram velocity to ram position; that is, it integrates from velocity to position. The integration is indicated by
$1/s$. This and other dynamic terms which are functions of s will
become clearer when we discuss Laplace transforms. Recall from
Chapter 1 that the human interface created the integration to position while also monitoring velocity.

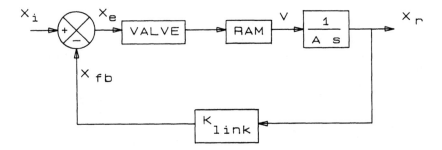

FIGURE 3.4 Position-control block diagram.

Block-diagram representation of a system becomes simplified
when analyzed in the s-domain rather than in the time domain.
The complexity of time-domain representations (convolution) is
reduced to multiplication in the s-domain. The result of using
the s-domain in block diagrams is that the overall transfer function is obtained from simple algebraic rules applied to the individual block diagrams of the loop. Each block of the loop represents a function of the loop, such as a spool valve, actuator dynamics, or feedback element.

Figure 3.5 shows the standard format of linear, single-input,
single-output systems. The plant function G(s) contains the total dynamics of the components and subsystems to be analyzed.
R(s) is the input command to the loop, and C(s) is the output
from the plant. Without the feedback H(s), the system is open

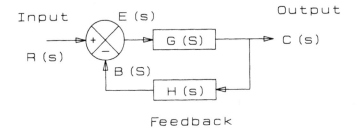

FIGURE 3.5 Block-diagram terminology in s-domain.

loop. The major problem with an open-loop system is its inability to maintain correlation between input and output in the presence of changes in load or other disturbances (such as temperature or supply pressure variations). By providing a feedback mechanism, such as the linkage in Figure 3.3, the output is maintained at the level commanded.

The transfer function of the block diagram of Figure 3.5 is a mathematical reduction of any single-input, single-output system; it allows for stability and controllability assessment of a system. Mathematically the transfer function can be derived as follows:

$$C(s) = G(s)E(s)$$

$$E(s) = R(s) - B(s) = R(s) - H(s)C(s)$$

$$C(s) = G(s)[R(s) - H(s)C(s)]$$

$$C(s) + G(s)H(s)C(s) = G(s)R(s)$$

$$C(s)[1 + G(s)H(s)] = G(s)R(s)$$

$$\frac{C(s)}{R(s)} = \frac{G(s)}{1 + G(s)H(s)} = \text{transfer function} = \text{T.F.}$$

The resulting transfer function relates an output $C(s)$ to the input $R(s)$ as a ratio of polynomials, with the plant dynamics in the numerator and denominator and the feedback elements incorporated in the denominator. Intuitively, if $G(s)H(s) \gg 1$, the transfer function reduces to $1/H(s)$, which is highly desirable (because the dynamics of the forward loop would not affect the output). Each dynamic representation is assumed to be in the form of a function of s. One needs to establish the dynamic terms of the plant, such as the valve and actuator, and maintain good, but stable, control in their interaction; the Laplace transform becomes the key to understanding, predicting, and accomplishing such control.

3.2 LAPLACE TRANSFORM

The Laplace transform is defined as

$$\mathcal{L}[f(t)] = F(s) = \int_0^\infty f(t)e^{-st}\, dt$$

where

    s   = complex variable

    $\mathcal{L}$ = symbol which indicates the Laplace operation

    F(s) = Laplace transform of the function f(t)

    The Laplace-transform representation of block diagrams permits easy analysis and solution of complex systems. This will become more apparent after a brief look at the convolution integral and some typical first- and second-order systems. A first-order system with a constant input A,

    y' + ay = A

has the exponential time-domain solution

    $f(t) = Ae^{-at}$ (A and a are constants)

This type of exponential term occurs often in physically decaying systems. This and the second-order system will be used extensively throughout this book. The Laplace transform of this first-order response is

$$\mathcal{L}[f(t)] = \mathcal{L}[Ae^{-at}] = \int_0^\infty (Ae^{-at})e^{-st}\,dt = A\int_0^\infty e^{-(a+s)t}\,dt$$

$$= \frac{Ae^{-(a+s)t}}{-(a+s)}\Big|_0^\infty = \frac{A}{s+a}$$

The integration process is equivalent to dividing by s (see Figure 3.4). This is proved as follows:

$$\mathcal{L}\left[\int f(t)\,dt\right] = \int_0^\infty \left[\int f(t)\,dt\right]e^{-st}\,dt$$

$$= \left[\int f(t)\,dt\right]e^{-st}\Big/(-s)\Big|_0^\infty - \left[\int_0^\infty f(t)\,dt\right]e^{-st}\Big/(-s)$$

$$= \frac{1}{s} \int f(t) \; dt \Big|_{t=0} + \frac{1}{s} \int_{0}^{\infty} f(t) e^{-st} \; dt$$

$$= \frac{f^{-1}(0)}{s} + \frac{F(s)}{s}$$

If the initial value of the integral is zero, the Laplace transform of the integral of $f(t)$ is $F(s)/s$. The Laplace transform of the derivative is

$$\mathcal{L}\left[\frac{df(t)}{dt}\right] = sF(s) - f(0)$$

where $f(0)$ is the value of $f(t)$ at $t = 0$. This is proved with an integration by parts:

$$\mathcal{L}\left[\frac{df(t)}{dt}\right] = \int_{0}^{\infty} e^{-st} \frac{df(t)}{dt} \; dt$$

$$= f(t) e^{-st} \Big|_{0}^{\infty} + s \int_{0}^{\infty} e^{-st} f(t) \; dt$$

$$= sF(s) - f(0)$$

where $f(0)$ is the initial condition.

If the initial value is zero, this becomes $sF(s)$ and the process of differentiation is equivalent to multiplying by s. Similarly, the second derivative of $f(t)$ is

$$\mathcal{L}\left[\frac{d^2 f(t)}{dt^2}\right] = s^2 F(s) - sf(0) - \dot{f}(0)$$

$$= s^2 F(s) \quad \text{when the initial conditions are zero}$$

where $f(0)$ is the value of $d/dt \{f(t)\}$ evaluated at $t = 0$.

Note the appearance of s^2 as a multiplier. Systems analysis with Laplace transforms often uses initial conditions of zero, with the input $R(s)$ dictating the form of this zero-state response. The convolution integral yields the zero-state (or particular) solution in the time domain. We will show that it is simplified and equivalent to multiplication in the s-domain. Figure 3.6 is a block

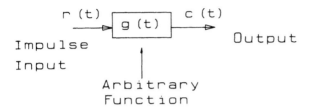

FIGURE 3.6 Impulse input.

diagram of a time-domain function g(t) acted on by an impulse input r(t) to produce the output c(t). Figure 3.7 is a plot of the
input r(t') versus t' upon the system g(t').

A differential impulse is shown in the shaded area with height
r(t') and width (dt'). This differential impulse is obtained by
scaling the unit impulse in Figure 3.8. The unit impulse may be
represented by an extremely tall, narrow pulse of unit area. The
unit-impulse response is g(t), which is the response at time t due
to the unit impulse at t' = 0. The area of the shaded portion of

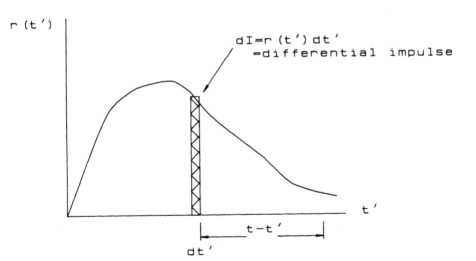

FIGURE 3.7 Impulse response. Area under curve (integral of
differential impulses) is the total response.

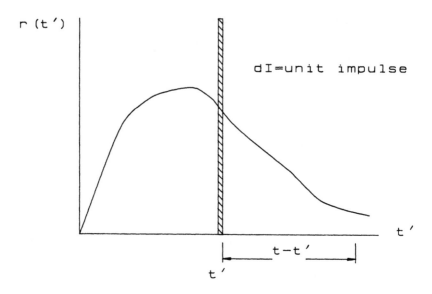

FIGURE 3.8 Unit impulse.

Figure 3.7 is $dI = r(t')\ dt'$. The response at t due to the differential impulse at $t' = 0$ is $g(t)\ dI$. Similarly, the response at t
due to the differential impulse at $t' \neq 0$ is $g(t - t')\ dI$. The "sum"
over a continuous sequence of such impulses gives

$$\int_0^t g(t - t')\ dI \quad \text{or} \quad c(t) = \int_0^t g(t - t')r(t')\ dt'$$

To get a better feel for the convolution, we present a simple system to portray its implementation. The first-order system discussed previously,

$$y' + ay = A \quad \text{with } y(0) = 0$$

can also be solved directly in the s-domain. The Laplace transform converts this differential equation into

$$sY(s) + aY(s) = A$$

so that

$$Y(s) = \frac{A}{s + a} = \frac{A}{s + 1/T} \qquad \text{where } T = \frac{1}{a}$$

As we have seen previously, this is equivalent to $y(t) = Ae^{-at}$.

Figure 3.9 shows the equivalence between convolution in the time domain and multiplication in the s-domain. The first-order example with a unit-step input will be used to show a typical result of this equality. The first-order exponential system is

$$g(t) = Ae^{-t/T}$$

and the unit step input is

$$r(t) = 0 \qquad \text{for } t < 0$$

$$= 1 \qquad \text{(a constant) for } t > 0$$

The Laplace transform of a step function is

$$\mathcal{L}[1] = \int_0^\infty 1e^{-st}\, dt = \frac{1}{s} \qquad \text{(see integral derivation)}$$

Therefore the unit step is 1 in the time domain and $1/s$ in the s-domain, or

$$r(t) = 1 \qquad \text{and} \qquad R(s) = \frac{1}{s}$$

In the time domain, the output is the convolution of the input with the unit-impulse response, or

$$c(t) = \int_0^\infty r(t')g(t - t')\, dt' = \int_0^t e^{-(t-t')/T}\, dt' = e^{-t/T} \int_0^t e^{t'/T}\, dt'$$

$$= e^{-t/T} Te^{t'/T}\Big|_0^t = e^{-t/T}\, \frac{e^{t/T} - 1}{1/T}$$

$$= T\left(1 - e^{-t/T}\right)$$

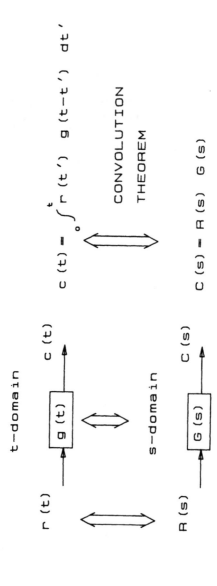

FIGURE 3.9 Convolution equivalence. Time-domain integral is simplified by s-domain multiplication.

In the s-domain, the equivalent procedure is

$$C(s) = G(s)R(s) = \frac{1}{s + 1/T} \quad \frac{1}{s} \quad = \frac{T}{s} + \frac{-T}{s + 1/T}$$

Therefore the inverse Laplace transform becomes

$$c(t) = AT\left(1 - e^{-t/T}\right)$$

which is equivalent to the time-domain solution. The time-domain and s-domain techniques represent, in a one-to-one fashion, equivalence among the input, the function, and the resulting output. The majority of systems analyzed in this book use the powerful, simple, s-domain method for system evaluation. Digital controls will be analyzed with the Z transform. Chapter 6 branches into the state-space analysis of systems. The block diagram (of Figure 3.5), shown and discussed previously, can be used for representing and evaluating linear control systems.

The block diagram plays a very powerful role in performance and stability predictions for many systems. A further study of the valve and actuator will portray the importance of this mathematical treatment of block-diagram modeling.

The dynamics of the valve-actuator combination of Figure 3.1 will be derived from the block diagram. Addition of the feedback link will show the mathematical results and an interpretation of the system. The spool position X_V of Figure 3.1 causes flow output to the actuator. This flow is also a function of the pressure differential across the valve. In a linearized equation, this valve output flow is

$$q = K_q X_v - K_{pq} \Delta P$$

where

K_q = flow gain of the valve (cubic inches per second per inch of stroke)

K_{pq} = slope of the load-flow curves at the operating point in Figure 3.10

ΔP = differential load pressure $(P_1 - P_2)$ across the ram

The flow q is equivalent to the velocity of the ram (created by this flow) multiplied by the pressurized area of the ram or

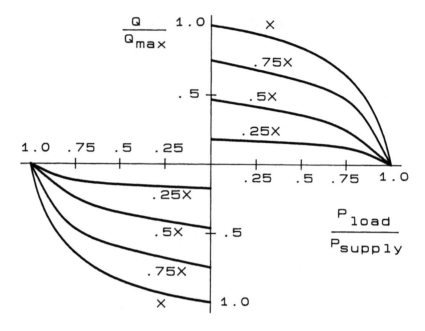

FIGURE 3.10 Load-flow curve. Slope at operating point is K_{pq}. Each curve is a function of the maximum stroke X.

$$Q = A\dot{X}_{ram} = K_q X_v - K_{pq} \Delta P$$

The force developed by the actuator is $A\Delta P$, or

$$F_a = \Delta PA = \frac{K_q X_v - AX_r}{K_{pq}}$$

Since there is no spring present, the ram moves the mass load (which will have viscous friction). From Newton's second law of motion, the sum of the forces on the mass is equal to the mass times its acceleration:

$$m\ddot{X}_r = -f\dot{X}_r + F_a$$

where

$f\dot{X}_r$ = viscous friction proportional to the velocity of the mass

F_a = input force on the mass

This is equivalent to

$$m\ddot{X}_r + f\dot{X}_r = \frac{K_q X_v - A X_r}{K_{pq}}$$

$$m\ddot{X}_r + f\dot{X}_r + \frac{A X_r}{K_{pq}} = \frac{K_q X_v}{K_{pq}}$$

$$m\ddot{X}_r + \left(\frac{f + A}{K_{pq}}\right)\dot{X}_r = \left(\frac{K_q}{K_{pq}}\right)X_v$$

$$m\dot{V}_r + \left(\frac{f + A}{K_{pq}}\right)V_r = \left(\frac{K_q}{K_{pq}}\right)X_v$$

In the s-domain, this equation reduces to

$$msV_r(s) + \left(\frac{f + A}{K_{pq}}\right)V_r(s) = \left(\frac{K_q}{K_{pq}}\right)X_v(s)$$

which simplifies to

$$V_r(s)\left[ms + \left(\frac{f + A}{K_{pq}}\right)\right] = \left(\frac{K_q}{K_{pq}}\right)X_v(s)$$

The ratio of output to input yields the transfer function of the valve:

$$\frac{V_r(s)}{X_v(s)} = \frac{K_q/K_{pq}}{ms + \left(f + A/K_{pq}\right)} = \frac{K_q/K_{pq}}{ms + f'}$$

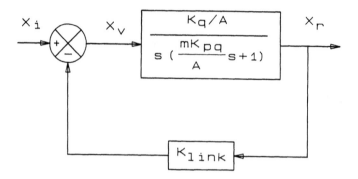

FIGURE 3.11 Position-feedback-control block diagram.

The term $f' = f + A/K_{pq}$ is an effective damping term. It is often desirable to write the transfer function in its normalized form. The normalized form is obtained by changing the dynamic elements (first-, second-, and higher-order terms) such that statically (slow inputs or $s = 0$) the term normalizes to a gain of unity. Normalizing the last transfer function gives the first-order time constant $T = m/f'$.

$$\frac{V_r(s)}{X_v(s)} = \frac{K_q/K_{pq}(f + A/K_{pq})}{m/\{f + A/K_{pq}\}s + 1} = \frac{K_q/(K_{pq}f + K_{pq}A/K_{pq})}{(m/f')s + 1}$$

$$= \frac{K_q/A}{(m/f')s + 1}$$

The final transfer function assumes the frictional damping is negligible in well-designed valves; this produces good load-flow curves. This transfer function is represented by the forward-loop (open-loop) dynamics shown in Figure 3.11 (excluding the integrator $1/s$). The integrator is created only when closing the loop. The output is ram velocity. The static gain resulting from the normalized dynamics is K_q/A. The result of the transfer function is a first-order lag characterized by the factor $1/(Ts + 1)$, which is an exponential decay system.

The larger the time constant, the slower the system will respond to input changes. This would easily occur with the more massive systems since the numerator of the time constant is proportional to m. Therefore, for a static input [slow changes where the dynamic lag (1/(Ts + 1) is negligible], the output velocity is the spool position multiplied by the flow gain divided by the area of the ram. Small values of K_{pq} (or horizontally profiled slopes of the load-flow curve) are desirable in reducing the time lag of the system. If the actuator were a motor, velocity output would be more meaningful. The velocity, changed to position by the linkage in Figure 3.3, allows the following equation to hold:

$$m\ddot{X}_r + \left(\frac{f + A}{K_{pq}}\right)\dot{X}_r = \left(\frac{K_q}{K_{pq}}\right)X_v$$

In the s-domain, this becomes

$$ms^2X_r(s) + \left(\frac{f + A}{K_{pq}}\right)sX_r(s) = \left(\frac{K_q}{K_{pq}}\right)X_v(s)$$

$$X_r(s)\left[ms^2 + \left(\frac{f + A}{K_{pq}}\right)s\right] = \left(\frac{K_q}{K_{pq}}\right)X_v(s)$$

$$G(s) = \frac{K_q/K_{pq}}{s(ms + f')} = \frac{K}{s(ms + f')}$$

The block diagram is shown in Figure 3.11, where

$$G(s) = \frac{K}{s(ms + f')} \quad \text{and} \quad H(s) = K_{link}$$

The transfer function of the closed-loop system is

$$T.F. = \frac{C(s)}{R(s)} = \frac{G(s)}{1 + G(s)H(s)} = \frac{K/s(ms + f')}{1 + \{KK_{link}\}/s(ms + f')}$$

$$= \frac{K}{s(ms + f') + KK_{link}} = \frac{K}{ms^2 + f's + KK_{link}}$$

The static gain or scale factor becomes

$$S.F. = \frac{1}{K_{link}}$$

Earlier it was stated that it would be desirable to have a transfer function which is $1/H(s)$ or the inverse of the gain of the feedback function. For the ram and actuator, the feedback gain is

$$H(s) = K_{link} = \frac{X_{fb}}{X_r}$$

It would be ideal if this were the total transfer function, because the scale factor then would be $1/K_{link}$ or

$$S.F. = \frac{X_r}{X_{fb}}$$

which is the proportionality desired. The scale factor is the overall static gain of a closed-loop component or system. However, the transfer function just derived has static gain $1/K_{link}$ altered by the dynamic second-order lag. The feedback linkage, in addition to changing the output from velocity to position, has allowed the system to be more tolerant to disturbances, such as pressure fluctuations and load-flow demands, while maintaining the desired output. The root-locus analysis is a mathematical method of studying these parameters on the complex plane, for predicting maximum gains with stability.

3.3 ROOT LOCUS

A further study of the dynamics of the open- and closed-loop systems is necessary for stability analysis and performance predictions. The complex plane is used to protray mathematically the significance of the transfer function. The complex variable is shown in Figure 3.12. A complex variable s has a real component $\zeta\omega_n$ and an imaginary component $j\omega$. A test point s_1 is

$$s_1 = \zeta\omega_{n_1} + j\omega_1$$

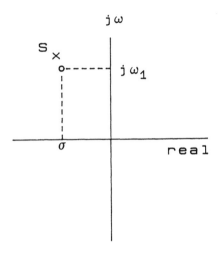

FIGURE 3.12 s-domain.

The complex plane is shown in Figure 3.13. The function G(s), itself a complex function, has real and imaginary parts

$$|G(s)| = G_x + jG_y$$

where the magnitude is $\sqrt{G_x^2 + G_y^2}$ and $\beta = \arctan(G_y/G_x)$. Points where the function or its derivatives approach infinity are called *poles*. For the valve-actuator system, $G(s)H(s) = K/s(Ts + 1)$ (the open-loop transfer function), there are poles at $s = 0$ and $s = -1/T$. Points where the function is zero are called *zeros*. In the transient response of a closed-loop system, the closed-loop poles provide the key characteristics. To obtain adequate performance yet maintain stability, we want to regulate the closed-loop pole and zero placement. We do this by first adjusting the complex-plane placement of the open-loop poles and zeros.

With more complex systems, it becomes difficult to directly calculate the closed-loop poles and zeros. The root-locus analysis lets us use a graphical technique to solve for the closed-loop poles and zeros with the open-loop variables of gain, poles, and zeros. The general closed-loop transfer function is

G (s) PLANE

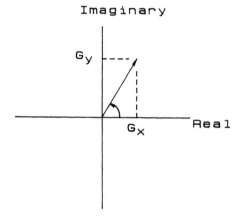

FIGURE 3.13 Imaginary plane.

$$\frac{C(s)}{R(s)} = \frac{G(s)}{1 + G(s)H(s)}$$

From the denominator, we obtain the "characteristic equation"

$$1 + G(s)H(s) = 0 \quad \text{or} \quad G(s)H(s) = -1$$

The G(s)-plane of Figure 3.13 contains the vector, in terms of magnitude and angle as

$$180°(2k + 1), \quad \text{where } k = 0, 1, 2, \ldots, \quad \text{and} \quad |G(s)H(s)| = 1$$

The s-plane values that satisfy these magnitude and angle conditions are the roots of the characteristic equation, or the system's closed-loop poles. The closed-loop block diagram of the valve-actuator system, as shown in Figure 3.14, has closed-loop transfer function

$$\text{T.F.} = \frac{K_0/s(Ts + 1)}{1 + K_0/s(Ts + 1)} = \frac{K_0}{s(Ts + 1) + K_0} + \frac{K_0}{Ts^2 + s + K_0}$$

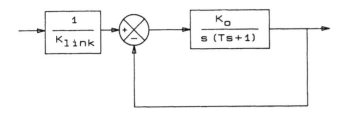

FIGURE 3.14 Closed-loop block diagram of valve-actuator system.

where $K_0 = K_q K_{link}/A$ is the open-loop gain. The characteristic equation is

$$Ts^2 + s + K_0 = 0$$

Figure 3.15 is a root-locus plot of the function. Note that the closed-loop poles at $K_0 = 0$ (labeled with the large X) are the same as the open-loop poles of $G(s)H(s)$, whereas the closed-loop zeros are factors of the zeros of $G(s)$ and poles of $G(s)H(s)$. The closed-loop poles on the real axis would be overdamped, or nonoscillatory. The point at which the plot just breaks away from the real axis corresponds to critical damping. As the gain and its resulting closed-loop pole enters the $j\omega$-axis, the closed-loop poles become complex and underdamped; that is, the system becomes oscillatory. The farther the pole is away from the origin, the smaller are the transients resulting from such a pole; they are also usually negligible in comparison to poles closer to the origin (if they exist).

The closer the poles are to the origin the more oscillatory the system will become. To ensure stability, choose no closed-loop poles in the right-half s-plane. If the system being studied were third order, the root-locus plot of Figure 3.16 would result. This system would not be stable if the gain were increased to its value at the intersection with the $j\omega$-axis. From a gain setting for a given system, it would be desirable to predict system reaction response to a change in input.

3.4 TIME RESPONSE

Transient and steady-state responses have been mentioned in terms of initial conditions of the Laplace transform and error

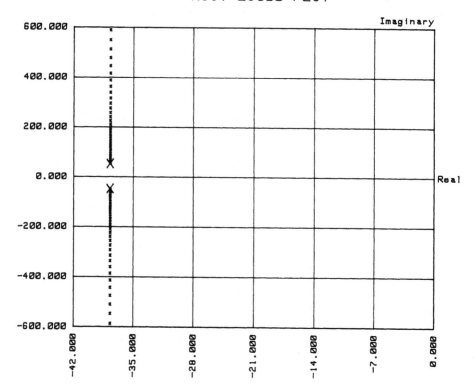

FIGURE 3.15 Root-locus plot of second-order system.

analysis. Transient response is the output which starts from the
initial state and goes to the final state. Steady-state response is
the output as t (time) approaches infinity. A system's stability
is directly related to its transient and steady-state responses.

If a control system remains at the same state in the absence
of disturbances, it is considered to be in equilibrium. A control
system is stable if, after a disturbance is applied, the system returns to its equilibrium state. The output may temporarily deviate from its steady-state conditions. This is not to be con-
fused with the droop or offset when there is no integration in
the system. Such an offset is a steady-state condition.

ROOT LOCUS PLOT

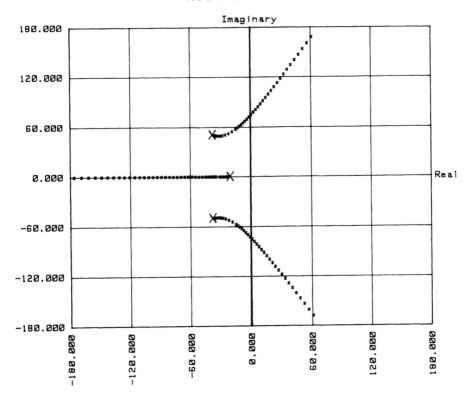

FIGURE 3.16 Third-order root-locus plot. Unstable for gain greater than 6.

3.4.1 First-Order Lag

The resulting transfer function of the valve actuator, without the linkage of Figure 3.1, is a first-order lag represented as

$$\text{T.F.} = \frac{C(s)}{R(s)} = \frac{1}{Ts + 1}$$

It is desirable to predict the behavior of the first-order system to a unit-step input. The unit step has Laplace transform $1/s$. The output becomes

$$C(s) = \frac{1}{Ts + 1} R(s) = \frac{1}{Ts + 1} \frac{1}{s}$$

which expands, by partial fractions, to

$$C(s) = \frac{1}{s} - \frac{1}{Ts + 1} = \frac{1}{s} - \frac{1}{s + 1/T}$$

This form of the output is in a configuration which can be transformed back in time (see Appendix 3). The resulting inverse Laplace transform becomes

$$c(t) = 1 - e^{-t/T}$$

At time $t = 0$ the output $c(t)$ is zero, and it eventually becomes unity at infinity. At time $t = T$ (the time constant of the system),

$$c(t) = 1 - e^{-1} = 0.632$$

or the output has reached 63% of its final value: the smaller the time constant T, the faster the system (or the less effect this lag has on the system). Figure 3.17 shows the time response of the first-order system subjected to the step input. For $t > 4T$, the output is at a steady state within 2% of the final value.

3.4.2 Second-Order Lag

The integration caused by the linkage feedback to this valve-actuator system (see Figure 3.14) raised the system to a second-order system with the general form

$$\text{T.F.} = \frac{C(s)}{R(s)} = \frac{C}{As^2 + Bs + C}$$

This can be factored (to show its poles) as

$$\frac{C(s)}{R(s)} = \frac{C/A}{\left[s + B/2A + \sqrt{(B/2A)^2 - C/A}\right]\left[s + B/2A - \sqrt{(B/2A)^2 - C/A}\right]}$$

If $B^2 - 4AC < 0$, the closed-loop poles are complex. For conventional reasons which will become obvious, the terms are redefined as

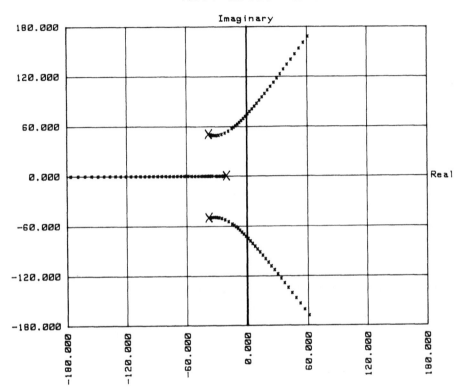

FIGURE 3.16 Third-order root-locus plot. Unstable for gain greater than 6.

3.4.1 First-Order Lag

The resulting transfer function of the valve actuator, without the linkage of Figure 3.1, is a first-order lag represented as

$$\text{T.F.} = \frac{C(s)}{R(s)} = \frac{1}{Ts + 1}$$

It is desirable to predict the behavior of the first-order system to a unit-step input. The unit step has Laplace transform $1/s$. The output becomes

$$C(s) = \frac{1}{Ts + 1} R(s) = \frac{1}{Ts + 1} \frac{1}{s}$$

which expands, by partial fractions, to

$$C(s) = \frac{1}{s} - \frac{1}{Ts + 1} = \frac{1}{s} - \frac{1}{s + 1/T}$$

This form of the output is in a configuration which can be transformed back in time (see Appendix 3). The resulting inverse Laplace transform becomes

$$c(t) = 1 - e^{-t/T}$$

At time $t = 0$ the output $c(t)$ is zero, and it eventually becomes unity at infinity. At time $t = T$ (the time constant of the system),

$$c(t) = 1 - e^{-1} = 0.632$$

or the output has reached 63% of its final value: the smaller the time constant T, the faster the system (or the less effect this lag has on the system). Figure 3.17 shows the time response of the first-order system subjected to the step input. For $t > 4T$, the output is at a steady state within 2% of the final value.

3.4.2 Second-Order Lag

The integration caused by the linkage feedback to this valve-actuator system (see Figure 3.14) raised the system to a second-order system with the general form

$$T.F. = \frac{C(s)}{R(s)} = \frac{C}{As^2 + Bs + C}$$

This can be factored (to show its poles) as

$$\frac{C(s)}{R(s)} = \frac{C/A}{\left[s + B/2A + \sqrt{(B/2A)^2 - C/A}\right]\left[s + B/2A - \sqrt{(B/2A)^2 - C/A}\right]}$$

If $B^2 - 4AC < 0$, the closed-loop poles are complex. For conventional reasons which will become obvious, the terms are redefined as

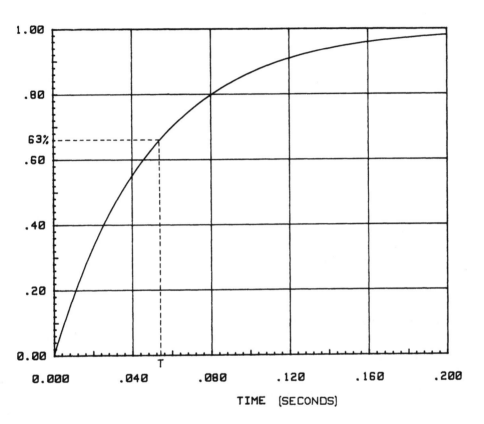

FIGURE 3.17 First-order time response.

$$\omega_n{}^2 = \frac{C}{A}, \qquad 2\zeta\omega_n = \frac{B}{A}$$

where

ω_n = undamped natural frequency

ζ = damping ratio of the system

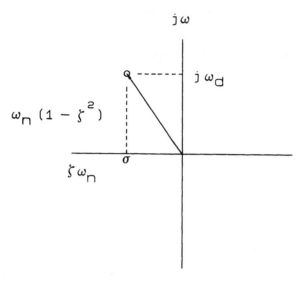

FIGURE 3.18 Second-order s-domain location.

Critical damping is obtained when

$$B_c = 2\sqrt{AC}$$

The damping ratio can also be defined from critical damping as

$$\zeta = \frac{B}{B_c} = \frac{B}{2\sqrt{AC}}$$

Figure 3.18 shows the relationship of these definitions with the complex plane. When $\zeta > 1$, $B = \sqrt{4AC}$ and the roots of the characteristic equation are real. The transfer function is now altered to

$$\frac{C(s)}{R(s)} = \frac{\omega_n^2}{s^2 + 2\zeta\omega_n s + \omega_n^2}$$

where $\omega_n^2 = C$, $2\zeta\omega_n = B$, $\zeta = B/2\omega_n = B/\sqrt{2C}$, and $A = 1$ which is factored as

$$\frac{C(s)}{R(s)} = \frac{\omega_n^2}{\left[s + \zeta\omega_n + \sqrt{(\zeta\omega_n)^2 - \omega_n^2}\right]\left[s + \zeta\omega_n - \sqrt{(\zeta\omega_n)^2 + \omega_n^2}\right]} -$$

$$= \frac{\omega_n^2}{\left[s + \zeta\omega_n + \omega_n\sqrt{\zeta^2 - 1}\right]\left[s + \zeta\omega_n - \omega_n\sqrt{\zeta^2 - 1}\right]}$$

Applications which contain second-order components are typically underdamped, with $0 < \zeta < 1$. Define the damped natural frequency as

$$W_d = \omega_n\sqrt{1 - \zeta^2}$$

The transfer function reduces to

$$\frac{C(s)}{R(s)} = \frac{\omega_n^2}{\{s + \zeta\omega_n + j\omega_d\}\{s + \zeta\omega_n - j\omega_d\}}$$

For a step input ($R(s) = 1/s$),

$$C(s) = \frac{\omega_n^2}{\left(s^2 + 2\zeta\omega_n s + \omega_n^2\right)s} = \frac{1}{s} - \frac{s + 2\zeta\omega_n}{s^2 + (2\zeta\omega_n)s + \omega_n^2}$$

Since $\omega_d = \omega_n^2(1 - \zeta^2) = \omega_n^2 - \omega_n^2\zeta^2$, the characteristic factor $s^2 + (2\zeta\omega_n)s + \omega_n^2$ can be rewritten as $(s + \zeta\omega_n)^2 + \omega_d^2$. The output due to the step input $1/s$ becomes

$$C(s) = \frac{1}{s} - \frac{s + \zeta\omega_n}{(s + \zeta\omega_n)^2 + \omega_d^2} - \frac{\zeta\omega_n}{(s + \zeta\omega_n)^2 + \omega_d^2}$$

From transform equivalence of Appendix 3, the inverse Laplace transform gives

$$c(t) = 1 - e^{-\zeta\omega_n t}\left\{\cos(\omega_d t) + \frac{\zeta}{\sqrt{1 - \zeta^2}}\sin(\omega_d t)\right\}$$

$$= 1 - \left(\frac{e^{-\zeta\omega_n t}}{\sqrt{1 - \zeta^2}}\right)\sin\left(\omega_d t + \arctan\sqrt{\frac{1 - \zeta^2}{\zeta}}\right)$$

UNIT STEP RESPONSE

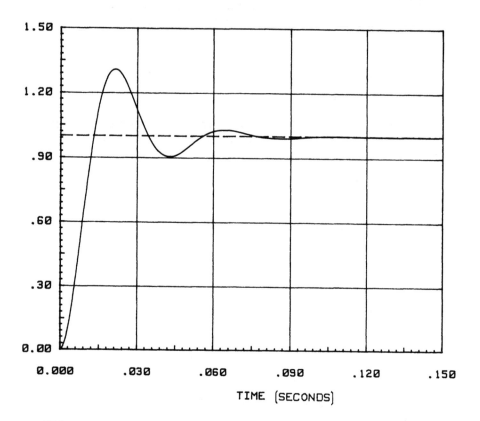

FIGURE 3.19 Second-order step response with $\zeta = 0.35$ and natural frequency of 25 Hz.

Figure 3.19 shows the response to a unit-step input, if $\zeta = 0.35$ and $\omega_n = 157$ rad/s. The frequency of oscillation is reduced (by the damping). The rise time (T_r), peak time (T_p), maximum overshoot, and settling time (T_s) basically define a step input's meaningful response parameters. The rise time is evaluated at $c(T_r) = 1$ or

$$c(T_r) = 1 = 1 - e^{-\zeta\omega_n T_r}\left\{\cos(\omega_d T_r) + \frac{\zeta}{\sqrt{1 - \zeta}}\sin(\omega_d T_r)\right\}$$

which reduces to

$$\tan(\omega_d T_r) = -\frac{\sqrt{1 - \zeta^2}}{\zeta} = \frac{\omega_d}{\zeta\omega_n}$$

When solved for T_r, this yields

$$T_r = \left(\frac{1}{\omega_d}\right)\arctan\left(\frac{\omega_d}{\zeta\omega_n}\right)$$

Similarly, the peak time is a function of ω_n and ζ. A small rise time is desirable. From Figure 3.18, it is evident that ω_n is inversely proportional to a small value of T_r. Therefore to keep T_r small, ω_n should be large. The damping will determine the stability about this effective rise time. The maximum overshoot caused by the damping should be kept within a certain range for stability and physical saturation reasons. The settling time is defined as a tolerance band within which the output will settle for a given step input. For a 2% tolerance band,

$$T_s = 4T = \frac{4}{\zeta\omega_n}$$

where T is the time constant of the system. Similarly, the larger the natural frequency (ω_n), the smaller the settling time. In addition, the damping has the same inverse effect. Therefore, larger damping is desirable for quick settling (unlike its role in rise and peak times). The compromise is to have sufficient damping for good settling, yet small enough to provide good rise time without excessive overshoot.

Since time response is the typical testing and operational mode for electrohydraulic systems, why should we need any other operating modes? Sinusoidal testing provides us with another method of analyzing a system. The root locus is good for relative stability and gain settings, and the time response gives us a feel for actual response time with stability limits. Frequency response, when properly used, provides a coherent means to evaluate response, stability, gain settings, and dynamic variable variations when components and systems are combined.

3.5 FREQUENCY RESPONSE

When electrohydraulic components are combined with other hydromechanical equipment, or when interfaced with electrical drives
and feedback, the complexity increases. Time-response calculations become very tedious (except in state-space analysis, which
is discussed in Chapter 6). Even though time response is infor-
mative, it becomes more difficult to single out problem areas or
to predict adequate compensation for higher-order systems.

If a sinusoidal input is used, and if its frequency is varied
over a prescribed range, the output spectrum is called the frequency response. The ability of a system to follow the sine input is a measure of the "goodness" of a system. The frequency-
response technique gives a better insight to the role of each component or element of a system.

The frequency response (Bode plot) provides the transfer
function of a system or component by applying a sinusoidal input and comparing it to the output (recall that the transfer function is the output divided by the input). The same spectrum may
be obtained with a dual-channel FFT (fast Fourier transform) ana-
lyzer. Mathematically, the transfer function is obtained by replacing s with $j\omega$. The block diagram of Figure 3.20 shows a
transfer function G(s), which represents a system of any order.
If the input r(t) is sinusoidal, $r(t) = A \sin(\omega t)$, the output is
also sinusoidal with a frequency-dependent phase and amplitude.
It will be shown that the transfer function C(s)/R(s) can be evaluated directly from G(s), where s is replaced by $j\omega$. The system
being studied, G(s), can be factored in terms of its poles and
zeros. For example, the second-order lag of the valve actuator
with linkage system is represented by the transfer function

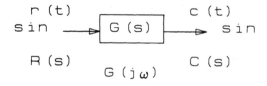

FIGURE 3.20 Sinusoidal input s-domain equivalence in block-diagram form.

$$G(s) = \frac{K}{(s + p_1)(s + p_2)}$$

where p_1 and p_2 are the poles or roots of the characteristic equation for second order. For a sinusoidal input, the Laplace transform is (see Appendix 3)

$$R(s) = \frac{\omega A}{s^2 + \omega^2}$$

The output becomes

$$C(s) = R(s)G(s) = \frac{\omega A}{s^2 + \omega^2} \frac{K}{(s + p_1)(s + p_2)}$$

$$= \frac{\tilde{n}}{s + j\omega} + \frac{n}{s - j\omega} + \frac{a}{s + p_1} + \frac{b}{s + p_2}$$

where a, b, $\tilde{n}$, and n are constants, and $\tilde{n}$ is the complex conjugate of n. The inverse Laplace transform gives

$$c(t) = ne^{-j\omega t} + \tilde{n}e^{j\omega t} + ae^{-p_1 t} + be^{-p_2 t}$$

Since we are looking for steady-state response (the output when t approaches infinity), and since $-p_1$ and $-p_2$ have negative real parts (for stability), the exponential terms involving p_1 and p_2 approach zero. Therefore the steady-state response reduces to

$$c(t) = ne^{-j\omega t} + \tilde{n}e^{j\omega t}$$

where n and $\tilde{n}$ are evaluated as follows:

$$\frac{\omega A}{s^2 + \omega^2} = \frac{n}{s + j\omega} + \frac{\tilde{n}}{s - j\omega}$$

$$\frac{\omega A}{s^2 + \omega^2}(s + j\omega) = n + \tilde{n}\frac{s + j\omega}{s - j\omega}$$

$$n = G(s)\frac{\omega A}{s^2 + \omega^2}(s + j\omega)\Big|_{s = -j\omega} = -\frac{AG(-j\omega)}{2j}$$

$$\frac{\omega A}{s^2 + \omega^2} (s - j\omega) = n \frac{s - j\omega}{s + j\omega} + \tilde{n}$$

$$\tilde{n} = G(s) \frac{\omega A}{s^2 + \omega^2} (s - j\omega)\Big|_{s = j\omega} = \frac{AG(j\omega)}{2j}$$

$$G(j\omega) = |G(j\omega)| e^{j\phi}$$

$$G(-j\omega) = |G(-j\omega)| e^{-j\phi} = |G(j\omega)| e^{-j\phi}$$

$$\phi = \arctan\left(\frac{G_j(j\omega)}{G_r(j\omega)}\right)$$

where G_j is the imaginary component and G_r is the real component. These constants result in the time-domain output

$$c(t) = \left\{\frac{AG(-j\omega)}{2j}\right\} e^{-j(\omega t + \phi)} + \left\{\frac{AG(j\omega)s}{2j}\right\} e^{j(\omega t + \phi)}$$

$$= A|G(j\omega)| \frac{e^{j(\omega t + \phi)} - e^{-j(\omega t + \phi)}}{2j}$$

$$= A|G(j\omega)| \left\{\frac{\cos(\omega t + \phi) + j \sin(\omega t + \phi)}{2j}\right.$$

$$\left. - \frac{\cos(\omega t + \phi) - j \sin(\omega t + \phi)}{2j}\right\}$$

$$= A|G(j\omega)| \sin(\omega t + \phi)$$

$$= A_1 \sin(\omega t + \phi) \quad \text{where } A_1 = A|G(j\omega)|$$

The system $G(s)$ with sinusoidal input will thus produce a sinusoidal output with the same frequency with frequency-dependent amplitude and phase. The frequency response of the system can be obtained directly from the system transfer function $G(s)$ by direct substitution of $j\omega$ for s, according to

$$G(s) \rightarrow G(j\omega) = \frac{C(j\omega)}{R(j\omega)}$$

The valve actuator and its version with linkage feedback represent typical first- and second-order lags. The first-order lag has transfer function

$$G(s) = \frac{K}{Ts + 1}$$

Substituting $j\omega$ for s, we get

$$G(j\omega) = \frac{K}{j\omega T + 1} \quad \text{where} \quad |G(j\omega)| = \frac{K}{\sqrt{\omega^2 T^2 + 1}}, \quad \phi = -\arctan(T\omega)$$

Therefore the first-order lag sinusoidal transfer function gives the sinusoidal output

$$c(t) = \frac{AK}{\{\sqrt{T^2\omega^2 + 1}\} \sin\{\omega t - \arctan(T\omega)\}}$$

If the input frequency range is very small, the dynamic influence of the lag is negligible and c(t) is equal to the gain of G(s) multiplied by the amplitude of the input (A).

The second-order lag system, such as the valve actuator with linkage feedback, may be represented as

$$G(s) = \frac{K}{s^2 + (2\zeta\omega_n)s + \omega_n^2} = \frac{K}{(s/\omega_n)^2 + (2\zeta/\omega_n)s + 1}$$

The sinusoidal frequency response is again obtained by substituting $j\omega$ for s:

$$G(j\omega) = \frac{K}{(j\omega/\omega_n)^2 + (2\zeta/\omega_n)j\omega + 1} = \frac{K}{j^2(\omega/\omega_n)^2 + (2\zeta\omega/\omega_n)j + 1}$$

$$= \frac{K}{\{1 - (\omega/\omega_n)^2\} + (2\zeta\omega/\omega_n)j}$$

The magnitude and phase angle become

$$|G(j\omega)| = \frac{K}{\sqrt{\{1 - (\omega/\omega_n)^2\}^2 + (2\zeta\omega/\omega_n)^2}}$$

$$\phi = -\arctan\left\{\frac{2\zeta\omega/\omega_n}{1 - (\omega/\omega_n)^2}\right\}$$

A linear plot of magnitude versus frequency becomes difficult to analyze. If the response is plotted on semilog paper with frequency on the log scale and decibels as the magnitude, it is easier to read and analyze. The decibel (dB) is defined as

M dB = 20 log K

where

M = magnitude in dB

K = gain of $G(j\omega)$

For example, if the magnitude is -6 dB, the corresponding gain K is 0.5. Typically, the frequency range of interest spans several decades. Over this range, the semilog plot retains the necessary information on a single plot. Figure 3.21 is a plot of the integrator, where the magnitude and phase are

$$M = 20 \log \left\{ \frac{1}{j\omega} \right\} dB = -20 \log |j\omega| \ dB = -20 \log(\omega) \ dB$$

$$\phi = -90°$$

The first-order lag gives the magnitude

$$M = 20 \log \left\{ \frac{1}{\sqrt{T^2\omega^2 + 1}} \right\} dB = -20 \log \left\{ \sqrt{T^2\omega^2 + 1} \right\}$$

as shown in Figure 3.22. The second-order lag yields

$$M = 20 \log \left\{ \frac{1}{\sqrt{1 - (\omega/\omega_n)^2 + (2\zeta\omega/\omega_n)^2}} \right\}$$

$$= -20 \log \left\{ \sqrt{1 - (\omega/\omega_n)^2 + (2\zeta\omega/\omega_n)^2} \right\}$$

which is shown in Figure 3.23 (with ζ = 0.3 and 0.7 and ω_n = $(2\pi)30$ = 188 rad/s). Note the correlation between this second-order frequency response and its equivalent time response: the larger the natural frequency ω_n, the quicker the rise time (of

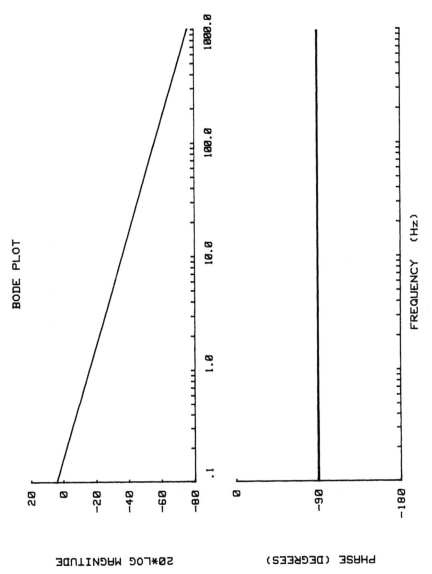

FIGURE 3.21 Integrator frequency response.

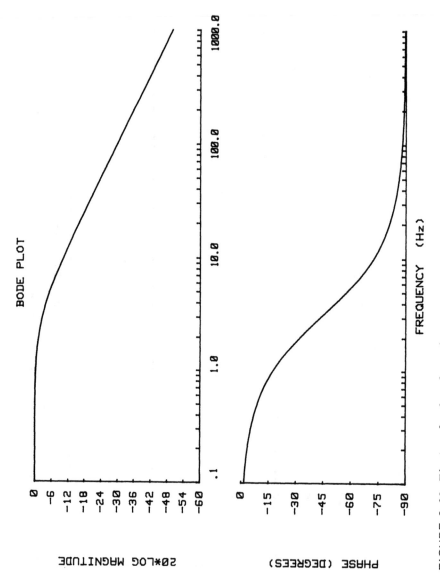

FIGURE 3.22 First-order lag dynamics.

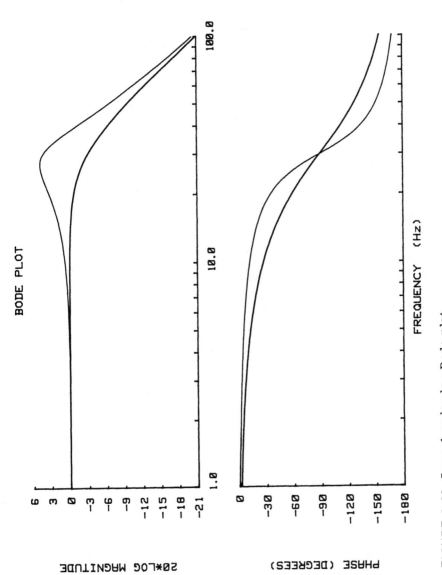

FIGURE 3.23 Second-order lag Bode plot.

its step response) and the larger the bandwidth (of the frequency response). The bandwidth is typically defined by either the 90° phase-lag frequency or the -3-dB magnitude frequency.

Stability is essential. The root-locus and step-response methods can be used to test for absolute stability. However, changes in (or addition of) dynamic terms aren't always obvious for higher-order systems. Stability predictions and compensating methods are clearer in the frequency plots.

The Nyquist stability technique is a method for obtaining the closed-loop stability from the open-loop frequency response. The open loop may employ actual Bode (frequency) plots of a component (such as the valve actuator) or it could give theoretical predictions. Chapters 5 and 6 use this stability technique for combining elements of electrical controllers, valves, pumps, and motors prior to any actual construction.

The Nyquist stability relates the quantity of open-loop zeros to the quantity of open-loop poles in the right half of the s-plane. Its derivation is based on complex variables and mappings on the s- and $F(s)$-planes, where $F(s) = 1 + G(s)H(s)$. One of the results of this stability criterion is the closeness of, or the margin of stability for, the open-loop frequency response.

The phase and gain margins of the open-loop frequency response relates the relative stability to key regions of the Bode plot. The positive and negative phase margins are shown in Figure 3.24; these margins are the amounts of phase required for relative stability. The frequency at which the magnitude of the open-loop transfer function is unity (or zero dB) is called the gain crossover frequency. The phase margin, also shown in Figure 3.24, is

$$\gamma_m = 180° + \phi$$

where

γ_m = phase margin

ϕ = phase lag of the open-loop transfer function at the gain crossover frequency

The gain margin K_{gm} is

$$K_{gm} = \left| \frac{1}{G(j\omega_x)} \right|$$

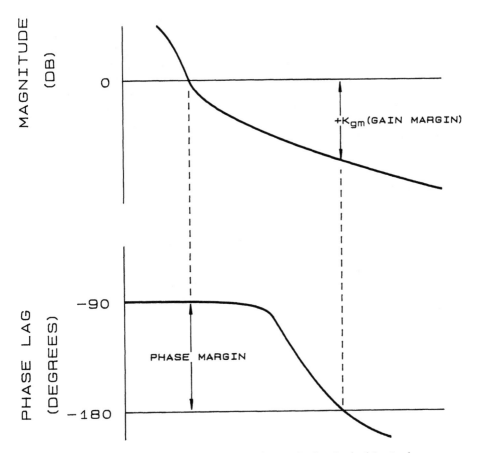

FIGURE 3.24 Phase and gain margins. It is desirable to keep
(gain margin) K_g > 6 dB and 30° < γ < 60° where (phase margin)
γ = 180° + ϕ (where ϕ is the open-loop phase angle of the system).

and

$$M_{gm} = 20 \log K_{gm} = -20 \log |G(j\omega_x)|$$

where

   M_{gm} = gain margin in dBs

   ω_x = frequency where the phase angle ϕ of the open loop is
           -180°

The first- and second-order lags are inherently stable. The
first- and second-order lags of the valve-actuator examples were
derived assuming other parameters (such as the fluid's bulk modulus and volume of oil) to be negligible. If the flow gain were
increased to an extreme, it could bring out these disregarded
components, resulting in a third- or higher-order system which
could be unstable. Chapter 5 discusses examples of higher-order systems and their stability, as well as compensating techniques to keep them stable or place them into stability.

3.6 ERROR OPTIMIZATION

The transient response to a step input and the dynamic bandwidth associated with a system are effective measuring sticks of
the system behavior. When the inputs vary from step to ramp
to sinusoidal and random noise, we want to know more about the
error signal generated by the system and how to optimize the error through the system parameters. Figure 3.25 is the standard
block diagram of a system with forward-loop dynamics $G(s)$ and
feedback elements $H(s)$. The actuating error signal becomes

$$\frac{C(s)}{R(s)} = \frac{G(s)}{1 + G(s)H(s)}$$

$$E(s) = R(s) - B(s)$$

$$B(s) = C(s)H(s)$$

$$C(s) = E(s)G(s)$$

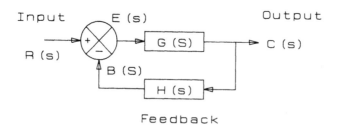

FIGURE 3.25 General block-diagram notation.

$$E(s) = R(s) - C(s)H(s) = R(s) - E(s)G(s)H(s)$$

$$E(s)[1 + G(s)H(s)] = R(s)$$

$$E(s) = \frac{1}{1 + G(s)H(s)} R(s)$$

From the final value theorem the steady-state actuating error E_{ss} becomes

$$E_{ss} = \lim_{t \to \infty} e(t) = \lim_{s \to 0} \frac{sR(s)}{1 + G(s)H(s)}$$

For a unit-step input, $R(s) = 1/s$, and the steady-state actuating error becomes

$$E_{ss} = \lim_{s \to 0} \frac{s}{1 + G(s)H(s)} \frac{1}{s} = \frac{1}{1 + G(0)H(0)}$$

For the system of Figure 3.25 (assuming an integrator in $G(s)$),

$$E_{ss} = \frac{1}{1 + K_0/0} = 0$$

If there were no integrator, the steady-state actuating error would be

$$E_{ss} = \frac{1}{1 + K_0}$$

In order to keep the offset, or steady-state error, to a minimum, the open-loop gain K_0 must be kept large. If the input is a ramp function,

$$E_{ss} = \lim_{s \to 0} \frac{s}{1 + G(s)H(s)} \frac{1}{s^2} = \lim_{s \to 0} \frac{1}{sG(s)H(s)}$$

For this same system with the integrator,

$$E_{ss} = \lim_{s \to 0} \frac{1}{sK_0(T_1s + 1)/s(T_2s + 1)(T_3s + 1)} = \frac{1}{K_0}$$

If the loop is a proportional loop (no integrator or 1/s),

$$E_{ss} = \lim_{s \to 0} \frac{1}{sK_0(T_1s + 1)/(T_2s + 1)(T_3s + 1)} = \frac{1}{0} = \infty$$

Therefore the system with integration will have an error when subjected to a ramp input. A large value of K_0 is desirable to keep the error low. Without integration, the system cannot maintain the relationship required of the ramp.

A system can be, and usually is, configured to obtain performance goals in terms of static and dynamic characteristics. A servovalve is specified to meet certain static characteristics in terms of hysteresis, linearity, and symmetry in its flow-gain and pressure-rise plots. Dynamically, it is designed to react to a step input within certain bounds or, equivalently, to obtain a certain bandwidth without excessive peaking. A system can also be chosen by how it reacts and handles the error signal.

If a performance index is chosen to optimize an error signal under certain mathematical rules, which can be related to the parameters of the system, the system will obtain a response consistent with the desired index. Therefore, if ' 1 index is chosen to produce a similar effect to step-input requirements, that index must have a range with a minimum or a maximum. Consider the performance index used in sizing step inputs: $\int_0^T e(t)^2 \, dt$. When this index is kept to a minimum, the system is optimal. Since the index is the accumulated square of the error signal, and since it is kept to a minimum, large errors will be changed rapidly and small errors will be less important. The best adjustment of a system parameter must be distinctly different from a poor adjustment. In other words, the design criteria must produce obvious changes in output, as functions of an adjustment in a gain or dynamic term.

If a PID controller (discussed in Chapter 5) drives a system, wherein the derivative time constant can be tuned to optimize a performance index, it would be desirable to choose an index which could tell good from bad results. The integral-squared index is not as selective as would be desired, because it relies heavily on large errors and on the error history. If the index were instead $\int_0^T te(t) \, dt$, then at the start of a step input, where t is small, the index will be small, even for a large error. The integral of error squared will have a larger value at the start of a step input. As the system approaches the desired output, the time t is large. Since the goal is to minimize the index, the index will force a large error to become minimal at larger values of time. A system based on this performance index will tame any overshoots, thus settling the response with adequate damping.

For higher-order systems and systems with compensation techniques, it becomes difficult to derive the performance index within its minimal limits. Experimentally, the index can be obtained by plotting successive runs of the integrand te(t) with respect to time, for a step input. If the desired parameter(s) are changed, the plot with the minimal area under the curve, over the time period 0 to T, satisfies the minimum performance-index requirements.

For microprocessors used in electrohydraulic systems with low bandwidths (such as heavy equipment), various schemes can optimize the system with respect to its error signal. An index can change during a portion of a step input. The microprocessor can then adjust a parameter of the controller to fit the index. Digital methods are numerous, and the advantages reflect the system usage.

3.7 DIGITAL CONTROLS

The control schemes and electrohydraulic systems discussed have been analog systems in which all signals, electrical or hydraulic, are continuous functions of time. The microprocessor adds a new dimension for the controller and therefore for the total system. The controller uses a microprocessor chip, specialized for the application, which represents data as discrete values. This digital means of system control, depending on the application, can add considerable flexibility and advantages, or it can cause unacceptable performance. Cost depends on the simplicity of the system and the need for its implementation.

If a system can be fully accomplished by hydraulic means with no benefits established by digital means, the microprocessor could look very expensive. If, however, other functions (which may or may not be related to the microprocessor) could benefit from a microprocessor, then the cost can balance out with increased performance. If the all-hydraulic system involved several stages of valving compensation, the microprocessor could easily be justified. Performance from the microprocessor is keyed to the plant. The microprocessor, when performing multiple tasks, including closed-loop functions, may have cost as well as performance advantages.

A simple microprocessor may be adequate for visual displays
and other "open-loop" needs, but it can't react fast enough for
more demanding systems. Some dedicated systems are configured to respond to signals quickly while performing other tasks.
For any microprocessor, there is a limit where its ability to react to a system is insufficient for the plant and system requirements under closed-loop control. It will become obvious that the
microprocessor can handle a 10-Hz bandwidth plant much easier
than a 100-Hz bandwidth system in a closed-loop mode.

The analog system was analyzed by the Laplace transform,
which replaced combinations of differential equations with algebraic equations, analyzed in the s-plane and the frequency domain. Similarly, the digital system involves difference equations
which are conveniently analyzed with the Z transform, and the
corresponding frequency response is investigated in the W-plane.

The particular implementation of the difference equation within
the W-plane is discussed in Chapter 5. The appendix discusses
the basics of the microprocessors, especially a versatile, dedicated chip which is ideal for control loops. Chapter 6 builds
upon the appendix and the material presented in this section to
obtain a closed-loop digital control scheme.

3.7.1 The Z Transform and the Difference Equation

The Z transform in the digital realm is analogous to the Laplace
transform for analyzing an analog system. The Z transform allows us to study the actual digital software and hardware of a
controller, the digital equivalent of the analog plant, and the
interface between the plant and controller. A typical digital-
analog system is shown in Figure 3.26. Appendix 2 and Chapter 5 give more detail to the controller. To appreciate the Z
transform in the digital plants and controllers, a single block
will be shown first to indicate the Z-transform equivalent of the
time and complex domains. Next, a closed-loop containing individual blocks with and without samplers will be mathematically
defined in the Z-transfer-function notation. Finally, a simple
system will indicate the difference between the analog and digital time responses. Before proceeding to the block diagram representation, we investigate the Z transform and a first-order
sample-and-hold circuit.

In order for an analog system or plant to be combined with a
digital system (microprocessor elements), the analog system must
be converted to a form acceptable to the digital controller. The

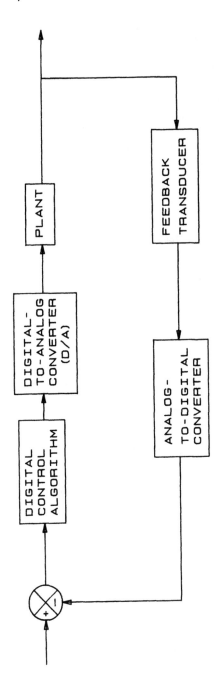

FIGURE 3.26 Analog-digital block diagram.

analog-to-digital (A/D) converter performs the matching from
analog to digital form through a sampler and holding device de-
picted in Figure 3.27. The sampler admits a value to the holding device of the A/D converter every T seconds, as set by the
microprocessor and converter. The holding device "holds" the
sampled value of the analog signal until the next sample is performed. The sampler behaves as a sequence of scaled impulses
at intervals of T seconds. When combined with the signal, the
sampling impulses carry the analog amplitudes at the sampling
instant shown in Figure 3.28. The sampled output x*(t) is

$$x^*(t) = \delta_T(t)x(t)$$

where $\delta_T(t)$ is a continuous set of unit impulses at sampling instances T. The unit impulse $\delta_T(t)$ is

$$\delta_T(t) = \sum_{k=0}^{\infty} \delta(t - kT)$$

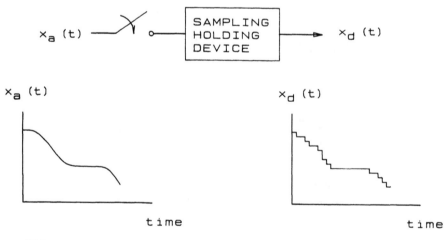

FIGURE 3.27 Sample-and-hold device.

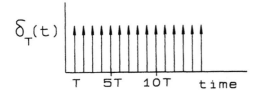

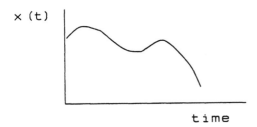

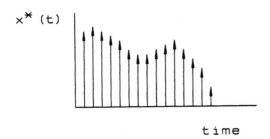

FIGURE 3.28 Unit-impulse sampler output from analog signal.

where $\delta(t - kT)$ evaluated at $t = kT$ is the unit-impulse function, or the sampling occurs only at multiples of T, and k is an integer. The sampler output becomes

$$x^*(t) = \delta_T(t)x(t) = \sum_{k=0}^{\infty} x(t)\delta(t - kT)$$

This is a form of the digital convolution in which the output $X^*(t)$ is the summation of the impulses (at the sampling instances kT) containing the heights of the analog signal. The Laplace transform of this sampled signal becomes

$$X^*(s) = \mathcal{L}[x^*(t)] = \sum_{k=0}^{\infty} x(kT)e^{-kTs}$$

Noting that the translation theorem states that $\mathcal{L}[\delta(t - kT)] = e^{-kTs}$, and by defining $z = e^{Ts}$, we can define the Z transform. Solving for s

$$\ln(z) = Ts, \qquad s = \frac{\ln(z)}{T}$$

and noting that $e^{-kTs} = (e^{Ts})^{-k} = z^{-k}$, we obtain for the Z transform

$$X(z) = X^*(s) = x^*\left(\frac{\ln(z)}{T}\right) = \sum_{k=0}^{\infty} x(kT)z^{-k}$$

which is the pulse transform. Thus, the pulse transfer function is the digital equivalent of the time function.

Since the Z transform depends on the analog signals at sampling multiples of T, the Z transforms of the analog signal and the sampled signal x*(t) contain the same information, or

$$Z[x(t)] = Z[x^*(t)] = X(z)$$

The unit-step function, represented as 1(t) in continuous form, is represented in the s-domain, through the Laplace transform, as 1/s. The Z transform of the unit step is

$$Z[x^*(t)] = Z[1(t)] = \sum_{k=0}^{\infty} 1(kT)z^{-k}$$

$$= 1 + z^{-1} + z^{-2} + \ldots = \frac{1}{1 - z^{-1}}$$

$$= \frac{z}{z - 1} \qquad \text{(unit step in Z form)}$$

The difference equation is an algebraic equation, in terms of Z, representing digital algorithms. Recall that the first- and second-order derivatives translated into the complex domain as

$$\mathcal{L}\left[\frac{d}{dt}\left\{f(t)\right\}\right] = sF(s) - f(0)$$

$$\mathcal{L}\left[\frac{d^2}{dt^2}\left\{f(t)\right\}\right] = s^2F(s) - s\dot{f}(0) - f(0)$$

Similarly, the Z transform of $x(k + 1)$ becomes

$$Z[x(k + 1)] = zX(z) - X(0)$$

where the notation has been simplified so that $X(k)$ represents $X(kT)$. If the initial condition were zero ($X(0) = 0$), then $zZ[x(k)]$ would represent a time shift of one period. Note that $z = e^{Ts}$ corresponds to the time domain as the unit impulse $\{x(t) = \delta(t - kT)\}$. Similarly, the second order becomes

$$Z[x(k + 2)] = zZ[x(k + 1)] - zx(1) = z^2X(z) - z^2X(0) - zx(1)$$

The exponent of Z is similar to the exponent of s for digital algorithms; it is equal to the value of the integer m for $x(k + m)$. The value z to the power 2 indicates a time shift of two periods.

The inverse Z transform Z^{-1} converts a Z transform into a time function. Several methods can be used to obtain the inverse. When the function of Z is not in a simple form corresponding to Table A.3, a technique such as partial fraction expansion is needed. Then the inverse Z transform is the sum of the partial-fraction terms which are reflected in Table A.3.

3.7.2 Holding Device

The holding device takes the information from the digital signal, represented in discrete levels at the sampling instants, and reproduces the signal in a continuous fashion. In order for the holding device to work properly, the sampling of the analog data must be at the correct rate in order to retain the necessary information of the plant. The holding device becomes a lowpass filter, which maintains the bandwidth of the plant and filters out higher-frequency noise. Shannon's sampling theorem states that a sampling frequency

$$W_s = \frac{2\pi}{T}$$

must be sized to the plant (signal) being sampled such that

$$W_s > 2W_{pm}$$

where W_{pm} is the maximum frequency of the plant which is indicative of its performance.

Naturally, the bandwidth must be within this region or all information will be lost. The limit is hard to define, especially if
the rolloff is gradual after the bandwidth is established. The
zero-order holding device is the most common; it holds the sampled signal as a constant until the next sample. Higher-order
sampling devices approximate the signal between two sampling
periods by polynomial terms. The transfer function of a zero-
order holding device is

$$G_h(s) = \frac{1 - e^{-Ts}}{s}$$

This can be shown by studying the combination of the sample-
hold pulses with their mathematical derivation. Figure 3.28 represents the waveform combinations which make up the sample-hold
circuit. Figure 3.29 represents $U(t - T) - u(t - 2T)$. Since

$$\mathcal{L}[\delta(t - kT)] = e^{-kTs}, \qquad \mathcal{L}[f(t - a)1(t - a)] = e^{-as}F(s)$$

the step at time k is

$$\mathcal{L}[U(t - kT)] = \int_0^{\infty} e^{-st} U(t - kT) \, dt = \int_0^{\infty} e^{-st}(0) \, dt$$

$$+ \int_{k+1}^{k} e^{-st}(1) \, dt = \frac{e^{-skT}}{a}$$

Similarly, the hold occurring between k and k + 1 is this step
function minus the k + 1 element:

$$\mathcal{L}[U(t - kT) - U(t - (k + 1)T)] = \mathcal{L}[U(t - kT)] - \mathcal{L}[U(t - (k + 1)T)]$$

$$= \frac{e^{-skT}}{s} - \frac{e^{-s(k+1)T}}{s}$$

$$= e^{-skT}\left(\frac{1 - e^{-sT}}{s}\right)$$

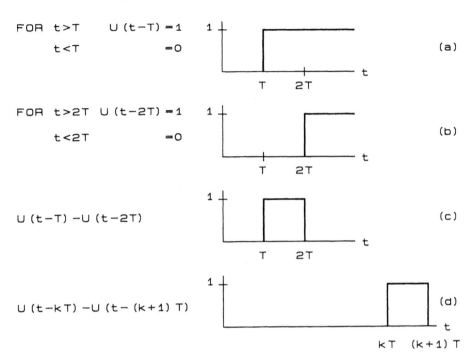

FIGURE 3.29 Sample-hold timing.

Figure 3.30 is a block diagram representing this resultant sample-hold circuit. The sample and hold can be used for both D/A and A/D converters. In the D/A, the sampler is effectively not used; theoretically, it doesn't have to be used in the A/D, except that it takes care of the fact that the A/D conversion isn't instantaneous.

3.7.3 Convolution and Pulse Transfer Functions

The pulse transfer function for a sampled plant G(s) can be derived from a convolution

$$y(kT) = \sum_{h=0}^{\infty} g(kT - hT)x(hT) = \sum_{h=0}^{\infty} x(kT - hT)g(hT)$$

FIGURE 3.30   s-domain block-diagram equivalent of sample-and-hold.

This reduces, through the Z transform, to

$$Y(z) = G(z)X(z)$$

where $G(z)$ can be shown to be

$$G(z) = \sum_{k=0}^{\infty} g(kT)z^{-k}$$

Thus the transfer function of a digital system can be evaluated directly from the impulse input. That is, once the transfer function (as a function of s) is known, the impulse response function $g(t)$, evaluated at $t = kT$, results in the pulse transfer function $G(z)$.

The first-order lag has the Z transform equivalent

$$G(s) = \frac{k_0}{s + a}$$

$$g(t) = K_0 e^{-at}$$

$$g(kT) = Ke^{-akT}$$

$$G(z) = \sum_{k=0}^{\infty} K_0 \, e^{-akT} z^{-k}$$

$$= K_0(1 + e^{-aT} z^{-1} + e^{-2aT} z^{-2} + \cdots)$$

$$= K_0 \frac{z}{z - e^{-aT}} = \frac{K_0}{1 - e^{-aT} z^{-1}}$$

When placing blocks together in the Z transform or pulse notation, note that the rules are strongly dependent on the location of the sampler. Whether the system has only two such blocks or contains a closed-loop arrangement of elements, the sampler location sets the style of implementing the Z transform.

The sampler typically occurs in the feedback path in electro-hydraulic systems employing a microcontroller. Obviously, two blocks in sequence vary in their Z transforms, depending on the placement of the sampler. If two blocks do not have a sampler between them, their Z transform is bound by first multiplying in the s-domain and then transforming into the Z-domain. Figure 3.31 is a typical closed-loop system with a sampler in the feedback. The sampler influences the derivation of the closed-loop block diagram as follows:

$$E(s) = R(s) - H^*(s)C(s)$$

$$C(s) = E(s)G(s)$$

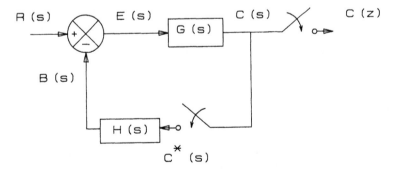

FIGURE 3.31 Sampled closed-loop block-diagram notation.

$$C(s) = E(s)G(s) = G(s)\{R(s) = H^*(s)C(s)\}$$

$$C(s) = G(s)R(s) - G(s)H^*(s)C(s)$$

$$C^*(s) = RG^*(s) - GH^*(s)C^*(s)$$

$$C^*(s)\{1 + GH^*(s)\} = RG^*(s)$$

$$C(s) = \frac{RG^*(s)}{1 + GH^*(s)}$$

$$C(z) = \frac{RG(z)}{1 + GH(z)}$$

In this notation, $RG(z)$ [and $GH(z)$] is the pulse transfer function of the Z transform of the blocks R and G [G and H] multiplied together. If the sampler was located at the output of the error signal, the closed-loop transfer function would be

$$C(z) = \frac{G(z)R(z)}{1 + GH(z)}$$

Note the difference between the two placements of the sampler. The first system requires multiplying the plant [G(s)] with the input [R(s)] before taking the Z transform, whereas the second system multiplies the individual Z transforms of the input [R(s)] and plant [G(s)]. These contrasts between the s- and Z-domains will become clear with a comparison of their responses.

3.7.4 Digital-Analog Equivalent

Figure 3.32 represents an analog plant driven by a digital controller. The arrangement is very typical for an electrohydraulic plant which interfaces with a digital controller. Note that the output $C(z)$ will be slightly different from that of the pulse transfer previously derived, due to the sampling at the input $R(s)$. The output is

$$C(z) = \frac{R(z)G(z)}{1 + GH(z)}$$

For this example, the digital compensator will be equal to unity. This gives us a picture of the dynamics involved for which compensation, through a digital controller algorithm, can be employed.

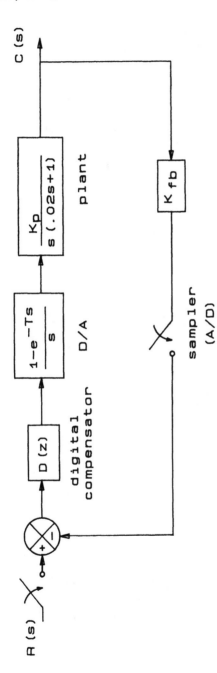

FIGURE 3.32 Analog-digital system.

Chapters 5 and 6 address the digital compensation in more detail.
The zero-order hold is a part of the plant G(s):

$$G(s) = \frac{1 - e^{-Ts}}{1 + GH(z)} = (1 - e^{-Ts})\frac{1}{s^2(0.02s + 1)}$$

The pulse transfer can then be obtained through the real translation (or shifting) theorem as

$$G(z) = Z\left[1 - e^{-Ts}\right]Z\left[\frac{1}{s^2(0.02s + 1)}\right]$$

$$Z\left[\frac{1}{s^2(Ts + 1)}\right] = T\left[\frac{T_s z}{(z - 1)^2} - \frac{(1 - e^{-Ts/T})Tz}{(z - 1)(z - e^{-Ts/T})}\right]$$

$$Z\left[1 - e^{-Ts}\right] = 1 - z^{-1} = 1 - \frac{1}{z} = \frac{z - 1}{z}$$

$$Z[s^2(0.02s + 1)] = T\left\{\frac{T_s z}{(z - 1)^2} - \frac{(1 - e^{-Ts/T})Tz}{(z - 1)(z - e^{-Ts/T})}\right\}$$

where

  T = 0.02 s

  T_s = sampling period

Therefore

$$G*(s) = G(z) = Z\left[1 - e^{-Ts}\right]Z\left[s^2(0.02s + 1)\right]$$

$$= K_x\left\{\frac{T_s}{z - 1} - \frac{(1 - e^{-Ts/T})T}{z - e^{-Ts/T}}\right\}$$

$$= K_x\left\{\frac{z\{T_s - T + e^{-Ts/T}\} - T_s e^{-Ts/T} + T - Te^{-Ts/T}}{z^2 - z\{1 + e^{-Ts/T}\} + e^{-Ts/T}}\right\}$$

where $K_x = K_0 T$

With the values of the loop defined, G(z) becomes

$$G(z) = \frac{0.6z + 0.0018}{z^2 - 1.61z + 0.61}$$

The closed-loop transfer function becomes

$$\frac{C(z)}{R(z)} = \frac{G(z)}{1 + G(z)} = \frac{K_x(0.6z + 0.0018)}{z^2 + (0.6K_x - 1.61)z + (0.0018K_x + 0.61)}$$

with the open-loop gain (K_0) set to a value of unity, the transfer function reduces to

$$\frac{C(z)}{R(z)} = \frac{0.6z + 0.0018}{z^2 - z + 0.61}$$

The system will be evaluated for a step input, in which

$$R(z) = \frac{z - 1}{z}$$

Therefore, the system response to the step input becomes

$$C(z) = \frac{z(0.6z + 0.0018)}{(z - 1)(z^2 - z + 0.61)}$$

$$= \frac{0.6z^2 + 0.00182}{z^3 - 2z^2 + 1.61z - 0.61}$$

By long division, this reduces to

$$
z - 2z + 1.61z - 0.61 \overline{\big)}
$$

$$
\begin{array}{l}
0.6z^{-1} + 1.2z^{-2} + 1.48z^{-3} + 1.4z^{-4} + 1.2z^{-5} + \cdots \\
\overline{0.6z^2 \quad + 0.0018z} \\
0.6z^2 \quad - 1.2z + 0.97 - 0.366z^{-1} \\
\quad\quad 1.2z - 0.97 + 0.37\ z^{-1} \\
\quad\quad 1.2z - 2.4\ + 1.93\ z^{-1} - 0.73z^{-2} \\
\quad\quad\quad\quad 1.48 - 1.56\ z^{-1} + 0.73z^{-2} \\
\quad\quad\quad\quad 1.48 - 2.96\ z^{-1} + 2.38z^{-2} - 0.903z^{-3} \\
\quad\quad\quad\quad\quad\quad 1.4\ \ z^{-1} - 1.65z^{-2} + 0.903z^{-3} \\
\quad\quad\quad\quad\quad\quad 1.4\ \ z^{-1} - 2.8\ z^{-2} + 2.25\ z^{-3} - 0.85z^{-4} \\
\quad\quad\quad\quad\quad\quad\quad\quad\quad\quad \overline{1.2\ z^{-2} + \cdots}
\end{array}
$$

Therefore, since the Z-domain transforms to the time domain through

$$Z[\delta(t - kTs)] = z^{-k}$$

the output, as a function of time, becomes the inverse Z transform at the sampling instances $c(kT_S)$.

Another way of establishing the same output is based on the difference equation of the transfer function. By rearranging the transfer function and transforming it into the time domain, a computer program can easily calculate the time response. The difference-equation approach is

$$\frac{C(z)}{R(z)} = \frac{K_x(0.6z + 0.0018)}{z^2 + (0.6K_x - 1.6)z + (0.0018K_x + 0.61)}$$

$$= \frac{K_x(0.6z^{-1} + 0.0018z^{-2})}{1 + (0.6K_x - 1.61)z^{-1} + (0.0018K_x + 0.61)z^{-2}}$$

$$C(z)\{1 + (0.6K_x - 1.61)z^{-1} + (0.0018K_x + 0.61)z^{-2}\}$$

$$= R(z)\{K_x(0.6z^{-1} + 0.0018z^{-2})\}$$

Therefore, the output becomes

$$c(kT_s) = 0.6K_x r[(k - 1)T_s] + 0.0018K_x r[(k - 2)T_s]$$

$$+ (1.61 - 0.6K_x)c[(k - 1)T_s] - (0.0018K_x + 0.6)c[(k - 2)T_s]$$

where $(k - 1)T_S$ represents the last value and $(k - 2)T_S$ represents the value before $(k - 1)T_S$.

A plot of the analog and digital systems is shown in Figure 3.33. Note that even though the plant is second order (which is inherently stable), it can become unstable in a digital system. Investigation of higher-order systems promptly leads to the conclusion that the s-to-Z transformation is cumbersome, to say the least. Several methods can be used to approximate the s-to-Z transform, one of which is shown in the following section.

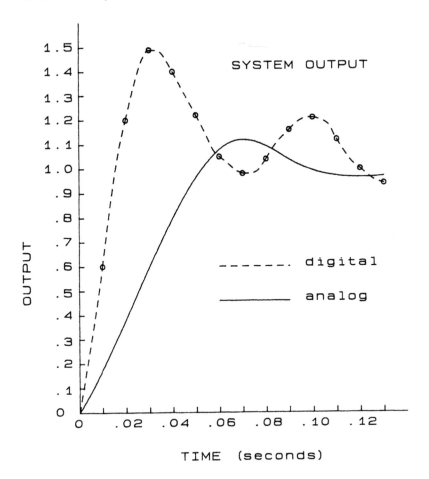

FIGURE 3.33 Analog-digital step response.

3.7.5 Bilinear Z Transform

Simulation of the integrator on a digital computer is not straightforward; neither is it easy to use the integrator to implement a
transformation between the s- and Z-domains. Many numerical techniques have been tried, with success rates dependent
upon the order of the system and stability of the integrator

method established. Chapters 5 and 6 use the *bilinear transfor-mation*. This will be introduced after first discussing several ap-proaches to digitally approximating the integrator. To obtain a better feel for the discrepancies, several methods will be shown for simulating the digital first-order system. The first-order lag was shown previously to be equivalent, by

$$\frac{C(s)}{R(s)} = G(s) = \frac{1}{Ts + 1}$$

$$G(z) = \frac{z}{z - e^{-Ts/T}} = \frac{1}{1 - e^{-Ts/T}z^{-1}}$$

Bibbero [2] implemented the first-order lag as

$$C_n = C_{n-1} + \frac{T_s}{T_s + T}(r - C_{n-1})$$

where

C_n = latest output value

C_{n-1} = last output value

T = time constant of first-order lag

T_s = sampling period of digital system

r = input to first-order lag

The derivation is

$$\frac{C(s)}{R(s)} = \frac{1}{Ts + 1}$$

$$TsC(s) + C(s) = R(s)$$

This is transformed into the time domain as

$$T\frac{dC(t)}{dt} + C(t) = r(t)$$

The derivative can be approximated by

$$\frac{dY}{dX}\Bigg|_{X = X_0} : \lim \frac{f(X_0 + \Delta X) - f(X_0)}{\Delta X} \simeq \frac{f(X_1) - f(X_0)}{X_1 - X_0}$$

where

$$X_1 - X_0 = \Delta T = T_s$$
$$f(X_1) = C_n$$
$$f(X_0) = C_{n-1}$$

The time-domain equation therefore reduces to

$$\frac{T(C_n - C_{n-1})}{T_s} + C_n = r$$

Solving for C_n, we obtain

$$C_n \left(\frac{T}{T_s} + 1 \right) = \left(\frac{T}{T_s} \right) C_{n-1} + r$$

$$C_n = \frac{TC_{n-1}}{T + T_s} + r \frac{T_s}{T + T_s}$$

$$= \frac{(T_s + T - T_s)}{T + T_s} C_{n-1} + \left\{ \frac{T_s}{T + T_s} \right\} r$$

$$= C_{n-1} + \frac{T_s}{T + T_s} \left(r - C_{n-1} \right)$$

It would be desirable to put this equation into a form which will fit with a lead circuit to form the digital equivalent of either a lead or a lag circuit. At the same time, it would be advantageous if it could be expressed in the Z-domain for easy implementation on the digital computer. The output C_n can be rewritten as

$$C_n = C_{n-1}(1 - a) + ar \quad \text{where } a = \frac{T_s}{T + T_s}, \; T_s = \text{sampling period}$$

$$\frac{C_n}{r} = \frac{C_{n-1}(1 - a)}{r} + a$$

Transforming this into the Z-domain, we obtain

$$\frac{C_n(z)}{r} = z^{-1}\frac{C_n(1 - a)}{r} + a$$

$$\frac{C_n(z)}{r(z)} \{1 - (1 - a)z^{-1}\} = a$$

$$\frac{C_n(z)}{r(z)} = \frac{a}{1 - (1 - a)z^{-1}} = \frac{a}{1 + Bz^{-1}} \quad \text{where } B = a - 1 = \frac{T}{T + T_s}$$

Note the similarity between this equation and the solution solved directly by the Z transform. This could be implemented as shown in Figure 3.34. Cushman [3] showed an integrator equivalent of

$$\frac{1}{s} = T_s \frac{z}{z - 1} = \frac{T_s}{1 - z^{-1}}$$

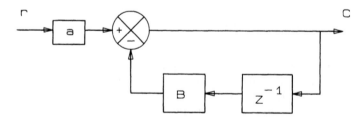

FIGURE 3.34 Digital lag compensation algorithm.

Since an integrator K_i/s becomes, within a closed loop, a first-order lag, $1/(Ts + 1)$ where $T = 1/K_i$, the Z-domain equivalent becomes (under closed-loop control) a first-order lag with a transfer function:

$$\frac{T_s/(1 - z^{-1})}{1 + T_s/(1 - z^{-1})} = \frac{T_s}{1 - z^{-1} + T_s} = \frac{1}{(T_s + 1)/T_s - (1/T_s)z^{-1}}$$

This reflects a first-order lag similar to that resulting from the direct Z transform and also from Bibbero's result (modified into the Z-domain). Cushman also states that, concerning the accuracy of the digital integration, the average of the last and present inputs relay a better approximation to the algorithm. Therefore, the summing becomes $(1 + z^{-1})/2r(z)$. This results in a more accurate transformation of the integrator:

$$\frac{1}{s} = \frac{T_s}{2} \frac{1 + z^{-1}}{1 - z^{-1}}$$

This is the bilinear transformation from the s- to the Z-domain through the integrator.

Because the form

$$\frac{C(z)}{r(z)} = \frac{1}{1 - Bz^{-1}}$$

is simpler to implement than other methods (faster and more efficient) with fewer stability problems, it will be used for actual digital implementation of first-order filters on the microcontroller. On the other hand, when simulating a plant to obtain its digital equivalent (in the Z-form), the bilinear transformation will be used. The plant transformation is more critical, especially when it is transformed into the W-plane (discussed in Chapter 5) for sizing of the filter parameters.

The Z-forms transform is a method which expands the direct implementation of the Z transform of the integrator to any power. It results in a one-to-one correspondence between each negative power of s, s^{-1}, s^{-2}, s^{-3}, ..., with an equivalent polynomial in Z. The following is a summary of a development [4] based on a power series expansion of $\ln(z)$. As shown previously in this chapter, the integrator is directly related to the Z-domain by

$$G(s) = \frac{1}{s} = \frac{T}{\ln(z)}$$

The $\ln(z)$ can be represented by the power series as

$$\ln(z) = 2\left\{\frac{z-1}{z+1} + \frac{1}{3}\left[\frac{z-1}{z+1}\right]^3 + \frac{1}{5}\left[\frac{z-1}{z+1}\right]^5 + \cdots\right\}$$

Therefore, the integrator becomes

$$\frac{1}{s} = \frac{T}{\ln(z)} = \frac{T/2}{U + 1/3\ U^3 + 1/5\ U^5 + \cdots}$$

which by synthetic division reduces to

$$\frac{1}{s} = \frac{T}{\ln(z)} = \frac{T}{2}\left[\frac{1}{U} - \frac{1}{3}U - \frac{4}{45}U^3 + \frac{44}{945}U^5 + \cdots\right]$$

For any power of the integrator, the equivalence is

$$\frac{1}{s^n} = \left(\frac{T}{2}\right)^n\left[\frac{1}{U} - \frac{1}{3}U - \frac{4}{45}U^3 + \frac{44}{945}U^5 + \cdots\right]$$

$$\frac{1}{s^n} = \frac{G_n(z^{-1})}{(1-z^{-1})^n} = G_n(z^{-1})$$

where $G_n(z^{-1})$ is the Z-form of s^{-n}. The value for $n = 1$ is

$$s^{-1} = \frac{T}{\ln(z)} = G_1(z^{-1}) = \frac{T}{2}\left[\frac{1}{u}\right] = \frac{T}{2}\left[\frac{1 + z^{-1}}{1 - z^{-1}}\right]$$

which is the same as the bilinear transformation. This allows a direct replacement between the s- and Z-domains. The bilinear transform and the Z-forms can be modified by a constant to minimize "prewarping"; this is a nonlinear effect of the transformation which affects the frequency response of the two systems. Since the electrohydraulic systems are generally low bandwidth, the bilinear and the Z-forms transformations are good representations; they will be used in Chapters 5 and 6 for evaluating digital filters and combined analog and digital systems.

Digital controls and analog systems, transformed into the realm of the digital system, can be investigated with time-response analysis. Sizing parameters is not straightforward because of the dissimilarity between the stability boundaries of the s and Z transforms. Extension of the digital system onto the W-plane produces a method of graphically portraying the system. This digital compensation and frequency-response analysis will be discussed in Chapter 5. Digital controllers are well suited for lower-bandwidth analog electrohydraulic systems.

3.8 CONCLUSION

The control of electrohydraulic systems should not be considered an art or a trial-and-error method. The control engineer uses the mathematics of control theory combined with the mathematical model of the system components. The control theory presented in this chapter complements the valving combinations of the following chapter. Control analysis, when combined with the hardware and optimized to the control needs, results in complete systems. These systems can confidently be made stable prior to the actual hardware build. The key to the electrohydraulic systems is the servovalve, since it precisely links the electronics to the hydraulic control.

BIBLIOGRAPHY

1. Ogata, Katsuhiko. *Modern Control Engineering*, Prentice-Hall, 1970.

2. Bibbero, Robert J. *Microprocessors in Instruments and Control*, Wiley, 1977.

3. Cushman, Robert H. Digital simulation techniques improve μP-system designs, *Electronic Design*, Jan. 7, 1981, 142–149.

4. Kuo, Benjamin C. *Digital Control Systems*, Holt, Rinehart and Winston, 1980.

4

The Control of Pressure and Flow

4.1 INTRODUCTION

The key to harnessing the power developed by the hydraulic
pumping elements and controlling it with the essential control algorithm is the servovalve. To obtain good simultaneous control
of high pressure (1000 to 5000 psi) and flow (2 to 200 gpm) in a
closed-loop mode, a multistage valve becomes necessary, because
single-stage valves cannot provide the necessary power with adequate response. These valves are intended primarily to control
pressure or flow. As a result of controlling flow (or pressure)
the pressure (or flow) will vary to fit the output loading conditions.

The servovalve is used in closed circuits as auxiliary functions such as controlling pump displacement. They are typically
used in open-circuit pumping arrangements for driving loads under force-, position-, and velocity-controlled outputs.

Some valves will droop (lose the desired output), whereas
others may be closer to the ideal requirements. The proportional
valve discussed in Chapter 2 is intended to produce output flow
proportional to spool stroke. During various loading conditions,
this proportionality can deviate by an unacceptable amount. The
servovalve typically has shorter strokes and valve spool laps
which are closer to line-to-line null lap conditions.

The proportional valves are generally used in open-loop systems; servovalves can be used in open- and closed-loop systems.
The closed-loop systems need the broader frequency response of
the servovalves. When the proportional valves become large, a
pilot stage is used to obtain fast, remote response. Closed-loop
systems employ pilot valves to match low-energy input signals to
the high-energy system components.

The pilot valves are typically used to drive spool diameters
from 1/4 in. to 2 in. The spools they drive are used in numerous applications. Flow-control valves are used more frequently
than pressure-control valves; however, pressure-control valving
is typically used in high-volume applications (dependent upon
market requirements). The pilot stages themselves vary in design and function. The pilot valves include a jet pipe, solenoid,
torque motor direct, torque motor nozzle flapper, and "voice-coil"
arrangements.

These pilot valves utilize a magnetic circuit together with
an electrical coil to activate the hydraulic circuit. There are
trade-offs between the various designs. The voice-coil style

is dynamically the quickest responding pilot stage, but it requires high amperage and is the most expensive. The various designs can be compared in terms of contamination sensitivity, input power levels, leakage flows, dynamic response, and static performance characteristics.

The most prevalent pilot used in closed-loop systems is the nozzle-flapper, torque-motor arrangement shown in Figure 4.1. The valve can be arranged to perform two different types of outputs with basically the same parts. Techniques for mating the pilot to a second stage will depend on function and assembly. Prior to studying pilot-boost interfacing, we will investigate the magnetic and hydraulic circuits.

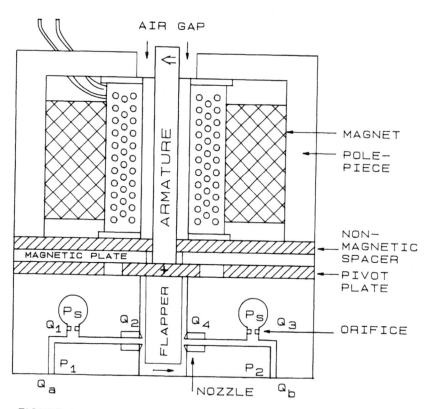

FIGURE 4.1 Nozzle-flapper pilot stage.

4.2 NOZZLE-FLAPPER PILOT VALVES

Nozzle-flapper pilot-stage valves exist in several configurations, depending on the magnetic circuit, feedback mechanism, and use within a system. System usage varies from static pressure and flow to dynamic response. The magnetic circuit provides the electrical interface to the hydraulic circuit.

4.2.1 Magnetic Circuit

The purpose of the magnetic circuit is to provide movement of the flapper at the nozzle, in order to activate the hydraulic circuit for a given electrical (or mechanical) input signal. The pivot plate also exists as a torsion tube. The armature is often shaped like a T with a coil on each branch. This T-bar-shaped armature changes the style of the magnetic circuit, but the function remains the same. The magnetic circuit has two effects when an electrical signal is fed into the coil. First, it provides a torque on the armature, to move it against the effective spring of the pivot plate or torsion tube. The resulting movement on the armature is reflected on the flapper through the pivot. This flapper movement, with respect to the nozzles, is the valve input to be controlled. Second, the magnetic circuit provides a decentering or negative spring rate. Sometimes, the circuit is designed to minimize the negative spring rate. In other cases, it is balanced to create a closed-loop pressure-control servovalve.

Since the pilot stage (used for this analysis) has a straight (rather than the T bar) armature, the magnetic circuit must provide a path throughout the length of the armature to the remainder of the circuit, in order to create movement at this armature. The magnetomotive force, or source energy to the magnetic circuit, is present at the air gaps of the flapper, as shown in Figure 4.1. The magnetomotive force F is analogous to the voltage source in an electric circuit. The electric circuit obeys Ohm's law

$$E = IR$$

where

E = circuit voltage

I = circuit current

R = circuit resistance

The magnetic circuit is analogous to the electric circuit, since it also has an "Ohm's law," given by

$$F = \phi R$$

where

F = magnetomotive force

ϕ = circuit flux

R = circuit reluctance through which the flux must flow

Figure 4.2 is a representation of the magnetic circuit. R_1 and R_2 are the main air-gap reluctances. At zero input current, $R_1 = R_2$, with equal air-gap distances producing no armature movement. The armature has a wire coil wrapped around it. The armature is shown originating at R_1 and R_2 and extending to the reluctance R_a. Through the armature branch of the circuit, the coil produces a magnetomotive input

$$F_c = N_c I_d$$

where

F_c = magnetomotive force due to coil

N_c = number of turns of wire on coil

I = differential current input

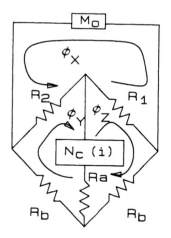

FIGURE 4.2 Magnetic circuit.

Reluctance R_a is shown because the armature must be part of the flux path, yet it must move. This causes an air gap at the bottom of the armature. The air gap is necessary because of the armature movement and because the pivot plate cannot be a magnetic material. Reluctances R_b (symmetric to the circuit) are present to balance the bridge. The circuit can be analyzed by the circuit fluxes ϕ_x, ϕ_y, and ϕ_z. The flux elements resulting at the top air gaps are the major concern. These air-gap fluxes are

$$\phi_1 = \phi_x + \phi_y, \qquad \phi_2 = \phi_x + \phi_z$$

In order to solve for the air-gap fluxes and relate them to the input torque on the armature, the reluctance of the variable gaps must be defined. The reluctance of a fixed air gap is

$$R_{gap} = R_g = \frac{G}{\mu A_g}$$

$$R_a = \frac{G_a}{\mu A_a}$$

$$R_b = \frac{G_b}{\mu A_b}$$

$$R_1 = \frac{G + x}{\mu A_g} = \frac{G + G[x/G]}{\mu A_g} = R_g \left(1 + \frac{x}{G}\right)$$

$$R_2 = \frac{G - x}{\mu A_g} = \frac{G - G[x/G]}{\mu A_g} = R_g \left(1 - \frac{x}{G}\right)$$

where

   μ = permeability of a particular portion of the magnetic circuit

   A = cross-sectional area of circuit parameter

   G = air-gap length

   x = linear position of armature from its neutral position

The reluctances R_1 and R_2 are functions of x (the travel of the armature with respect to its neutral position at the location of the top air gaps). With this definition, the circuit equations of Figure 4.2 reduce to the following equations:

$$\phi_x(R_1 + R_2) + \phi_y(R_2) \quad\quad + \phi_z(R_1) \quad\quad = M_0$$

$$\phi_x(R_2) \quad + \phi_y(R_2 + R_a + R_b) + \phi_z(-R_a) \quad\quad = N_c I$$

$$\phi_x(-R_1) \quad + \phi_y(-R_a) \quad\quad + \phi_z(R_a + R_b + R_1) = -N_c I$$

where

R_a = air gap at the bottom sides of the armature

R_b = spacer plate at the bottom of the magnetic circuit (see Figure 4.1)

When solved and placed back into the equations of the desired ϕ_1 and ϕ_2, the torque equation can be solved. The torque output (T) produced by the combination of the permanent magnet (M_0) and the coil, with its changing input current, is

$$T = \frac{4.42 \times 10^{-8} \, Gx}{(\mu A_g)} (\phi_2^2 - \phi_1^2) \, lb$$

This torque is reflected through the torque input (current) and the decentering motion, according to

$$T = K_{tm} I + K_m \alpha$$

where

K_{tm} = main torque motor gain (in.-lb/mA)

K_m = magnetic decentering spring rate (in.-lb/rad)

I = input current (mA)

α = angular position of armature

This torque equation is in the form which defines the parameters for open- and closed-loop block diagrams. The constants K_{tm} and K_m can be solved by nonlinear numerical methods (due to the

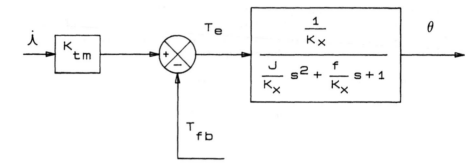

FIGURE 4.3 Block diagram of magnetic-circuit−flapper-pivot interface.

squared functions) combined with graphical techniques. Any coupling between x and I should be negligible for a well-designed circuit. With these terms defined, the torque output and feedback scaling can be evaluated.

4.2.2 Torque Summation

Figure 4.3 is a block diagram of the electromagnetic sections of the valve. The torque motor's influence on the system takes place before and inside the control loop. To see where the torque motor variables take effect and how they relate to the valve's output, we will derive both the input and output. The torque (T) summation is

$$\Sigma T = J\alpha = J\ddot{\theta}$$

$$J\theta = -f\ddot{\theta} - K_p\dot{\theta} - T_0 + T_m + (T_i - T_{fb})$$

$$= -f\theta - K_p\theta - K_0L_0^2\theta + K_m\theta + (T_i - T_{fb})$$

$$= -f\theta - (K_p - K_m + K_0L_0^2)\theta + (T_i - T_{fb})$$

$$= -f\dot{\theta} - K_x\theta + (T_i - T_{fb})$$

$$J\ddot{\theta} + f\dot{\theta} + K_x\theta = (T_i - T_{fb}) = T_e$$

where

θ = armature and flapper rotation

J = polar inertia of armature and flapper

f = velocity coefficient of friction

L_0 = pivot-to-nozzle distance

K_p = pivot stiffness

T_0 = torque due to hydraulic capacitance of pilot chambers

T_m = torque due to negative spring rate of magnetic circuit

K_m = magnetic spring rate (decentering)

T_i = input torque = iK_{tm}

T_{fb} = feedback torque

K_0 = oil spring rate (see Section 4.4)

The term K_x is a rotational spring rate which is affected by flow forces. These flow forces are discussed in [1]. This is Laplace transformed into the s-domain, resulting in

$$Js^2\theta(s) + fs\theta(s) + K_x\theta(s) + T_e(s)$$

$$\theta(s)[Js^2 + fs + K_x] = T_e(s)$$

$$\frac{\theta(s)}{T_e(s)} = \frac{1}{Js^2 + fs + K_x}$$

The block diagram is shown in Figure 4.3. The torque motor gain (K_{tm}) produces the torque input T_i as shown. The remaining torque gain (K_m) is a part of K_x, defined as

$$K_x = K_p - K_m + K_0L_0^2$$

where

K_p = stiffness of the torque armature pivot (in.-lb/rad)

K_m = magnetic decentering stiffness (in.-lb/rad)

$K_0L_0^2$ = effective stiffness of the oil in its chamber

Typically the magnetic decentering stiffness is sized to be much smaller than the stiffness of the pivot of the armature. The stiffness of the oil is usually made negligible by the pivot stiffness. A version of the nozzle-flapper valve which uses the decentering gain will be discussed after showing the feedback-wire nozzle-flapper valve.

4.2.3 Hydraulic Circuit

The hydraulic circuit of Figure 4.1 uses the position created by an electromagnetic circuit to create an unbalance. The porting is similar to that of a spool valve. At neutral position (no input armature movement), there are equivalent exit areas at the nozzles. With an orifice between the supply pressure and each output control port, an ambient pressure exists which is less than the supply pressure and greater than the tank pressure (because of the position of the nozzle with the flapper, which is itself an orifice).

Leakage exists at neutral. This leakage can be close to the actual output flow capability of the pilot stage. For a given input-position change of the flapper, say to the right (from a flapper movement to the left reflected through the pivot plate), the flow Q_4 will decrease because the orifice becomes smaller. This restriction to flow will cause the control port pressure P_2 to increase. Correspondingly, the drain flow Q_2 is less restricted; this lowers P_1. The output load-flow equations are

$$Q_a = Q_1 - Q_2 = C_{d0}A_0\sqrt{(2/\rho)(P_s - P_1)} -$$

$$C_{dn}\pi D_n(X_0 + X_e)\sqrt{(2/\rho)P_1}$$

$$Q_b = Q_3 - Q_4 = C_{d0}A_0\sqrt{(2/\rho)(P_s - P_2)} -$$

$$C_{dn}\pi D_n(X_0 - X_e)\sqrt{(2/\rho)P_2}$$

If the output control ports (C_a and C_b) are blocked,

$$Q_a = Q_b = 0$$

which reduces to

$$Q_1 = Q_2$$

$$C_{do}A_{ol}\sqrt{(2/\rho)(P_s - P_1)} = C_{dn}\pi D_n(X_0 - X_e)\sqrt{(2/\rho)P_1}$$

$$\frac{P_s - P_1}{P_1} = \left[\frac{C_{dn}\pi D_n(X_0 + X_e)}{C_{do}A_{ol}}\right]^2 = \frac{P_s}{P_1} - 1$$

$$P_1 = \frac{P_s}{1 + \left[\dfrac{C_{dn}\pi D_n(X_0 + X_e)}{C_{do}A_{ol}}\right]^2}$$

where

X_0 = null distance between flapper and nozzle

X_e = stroke of flapper at nozzle

Also,

$Q_3 - Q_4 = 0$

$$C_{do}A_{02}\sqrt{(2/\rho)(P_s - P_2)} = C_{dn}\pi D_n(X_0 - X_e)\sqrt{(2/\rho)P_2}$$

$$P_2 = \frac{P_s}{1 + \left[\dfrac{C_{dn}\pi D_n(X_0 - X_e)}{C_{do}A_{0_2}}\right]^2}$$

At null ($X_e = 0$), the pressures P_1 and P_2 are equal to the resulting ambient pressure at both control ports:

$$P_1 = P_2 = P_{null} = \frac{P_s}{1 + [C_{dn}\pi D_n X_0 / C_{do}A_0]^2}$$

where

D   = diameter

A_0 = area of orifice at the nozzle

o   = orifice

d   = discharge coefficient

n = nozzle

e = working stroke at the nozzle

The null ambient pressures can be set basically by the nozzle
and orifice diameters and the nozzle-flapper spacing, for a given
supply pressure. The nozzle-flapper spacing is typically 0.0015
to 0.003 in. for maximum performance. The discharge coefficients
(C_d) should be maximized, for true orifices, to minimize laminar
temperature effects. Depending on the method of valve interfacing, loading demands, and output pressure requirements, the
null pressure is set by the proper balance between the orifice
and nozzle diameters.

The primary objective of the pilot stage is to provide flow as
its output. The resulting pressure will vary with application,
unless it is a pressure-controlled pilot stage (which produces
an output differential pressure proportional to the pilot input).
This pressure-controlled pilot will be discussed later in the chapter. The flow gain of the pilot is the output flow created by the
nozzles for a given flapper deflection:

$$K_q = \frac{Q_1}{X_e}$$

The output flow for a given input current is shown in Figure 4.4.
The torque feedback at the summing junction can take on several
forms. If the boost stage fitted with the pilot is the typical spool
with feedback wire, the wire becomes the means of mechanically
providing the torque feedback onto the pilot stage (through the
stiffness of the pivot).

4.3 MULTIPLE-STAGE ELECTROHYDRAULIC
SERVOVALVES

Figure 4.5 is the nozzle-flapper pilot stage mated to a spool valve.
Input current, which produces armature deflection to the right,
will reflect a flapper movement to the left by means of the pivot.
This flapper position will create a higher pressure at the left nozzle than at the right nozzle, because of the restriction of flow to
tank pressure. This pressure buildup (P_1) at control port C_1

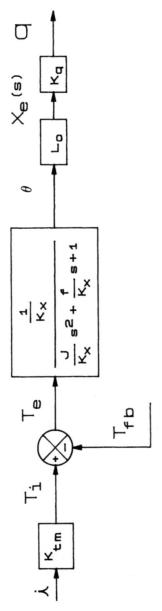

FIGURE 4.4 Pilot flow output block diagram.

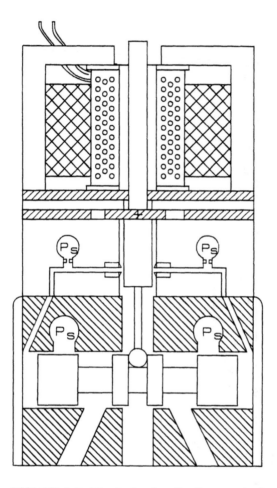

FIGURE 4.5 Electrohydraulic flow-control servovalve utilizing a
feedback wire for closed loop.

will divert the leakage flow to output flow at C_1. Velocity is a
result of the pilot flow working on the area of the spool:

$$V = \frac{Q_P}{A}$$

4.3.1 The Feedback Wire

Without the feedback wire, the spool would travel to a physical stop at each stroke. The wire will monitor the spool stroke and compare it to its desired position, thereby integrating the velocity (similar to the valve-actuated system discussed in Chapters 2 and 3). When the spool travels too far to the right, the wire will force the flapper to the right until an equilibrium exists at the flapper and armature.

If the wire overshoots the equilibrium point, the pressure will build up at the right-side control port; this will force the spool to the left and bring the feedback wire into a position which would retract the rightward flapper movement. This modulation is continued in a closed-loop position mode, to keep the spool position proportional to the input current. This spool position creates an output flow proportional to spool stroke, resulting in the block diagram of Figure 4.6.

The feedback-wire gain ($K_W = K_x l_0^2$) of Figure 4.6 can be sized to produce the necessary torque to match the pilot electromagnetic and mechanical input; this gives the proper scaling of nozzle-flapper displacement. The feedback wire can also be sized to produce the proper scale factor for the valve, which would determine the spool stroke for a given input signal. The pivot, flapper, and wire can be analyzed as separate sections to determine the proper sizing, as shown in Figure 4.7. For the beam (wire, flapper, or pivot), the deflection y with respect to its axial dimension x is

$$\frac{M}{EI} = \frac{d^2y}{dx^2}, \quad I = \frac{\pi D^4}{64}$$

where

y = wire deflection at any point x along its length

M = beam bending moment at x

E = elastic modulus of the wire

D = wire diameter

The first equation can be solved for M and integrated once for θ (the slope of the beam at x). Once θ is determined, it can be integrated to produce the deflection y:

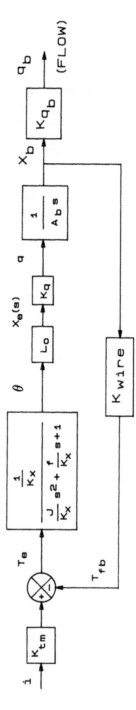

FIGURE 4.6 Flow-control servovalve block-diagram representation.

146

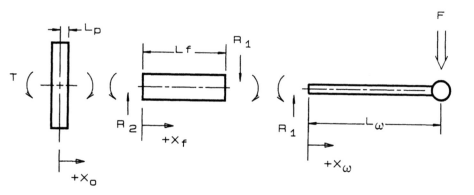

FIGURE 4.7 Feedback-wire sizing.

$$\theta = \int M \, dx = EI \frac{dy}{dx} + c$$

$$y = \int \theta \, dx$$

$$= \frac{1}{EI} \int \theta \, dx - c_1 x + c_2$$

For the wire (w subscripts) portion of the feedback, the bending moment is

$$M_w = E_w I_w \frac{d^2 Y_w}{dX_w^2} = M_1 - R_1 X_w$$

$$\theta E_w I_w = E_w I_w \frac{dY_w}{dX_w} = M_1 X_w - \frac{R_1 X_w^2}{2} + c_1$$

$$E_w I_w Y_w = \frac{M_1 X_w^2}{2} - \frac{R_1 X_w^3}{6} + c_1 X_w + c_2$$

The flapper (f subscripts) bending moment, slope, and displacement equations are

$$M_f = E_f I_f \frac{d^2 Y_f}{dX_f^2} = M_2 - R_2 X_f$$

$$E_f I_f \Theta = E_f I_f \frac{dY_f}{dX_f} = M_2 X_f - \frac{R_2 X_f^2}{2} + c_3$$

$$E_f I_f Y_f = \frac{M_2 X_f^2}{2} - \frac{R_2 X_f^3}{6} + c_3 X_f + c_4$$

The torque on the pivot, created by the angular position of the feedback, is

$$T = K_x \Theta = M_2$$

Solving for the boundary conditions one obtains

1. Θ_0 at $X_0 = L_{pf} = \Theta_a$ at $X_a = 0$ or

$$\frac{T}{K_x} = \frac{c_3}{E_f I_f}$$

2. Θ_f at $X_f = L_f = \Theta_w$ at $X_w = 0$ or

$$\frac{M_2 L_a - FL_a^2/2 + c_3}{E_f I_f} = \frac{c_1}{E_w I_w}$$

3. Y_0 at $X_0 = L_{pf} = Y_f$ at $X_f = 0$ or

$$\frac{\Theta L_{pf}}{2} = \frac{c_4}{E_f I_f}$$

4. Y_f at $X_f = L_p = Y_w$ at $W_x = 0$ or

$$\frac{M_2 L_f^2/2 - FL_f^3/6 + c_3 L_f + c_4}{E_f I_f} = \frac{c_2}{E_f I_f}$$

5. M_w at $X_w = 0 = M_f$ at $X_f = L_f$ or

$$M_2 - FL_f = M_1$$

6. M_a at $X_f = 0 = T = K_x \Theta$ or

$$M_2 = K_x \Theta$$

These equations can be rearranged to reveal a procedure which produces the unknowns in their proper sequence:

1. $M_1 = FL_w$

2. $M_2 = F(L_w + L_f)$

3. $\Theta = \dfrac{T}{K_x} = \dfrac{M_2}{K_x}$

4. $c_3 = \dfrac{M_2}{K_x} E_f I_f$

5. $c_1 = \dfrac{(M_2 L_f - FL_f^2/2 + c_3) E_w I_w}{E_f I_f}$

6. $c_4 = \Theta L_{pf} E_f I_f$

7. $c_2 = \dfrac{(M_2 L_f^2/2 - FL_f^3/6 + c_3 L_f + c_4) E_w I_w}{E_f I_f}$

Solving for the deflection of the wire (Y_w) and the deflection at the nozzle ($Y_{nozzle} = X_e$), we get

$$Y_w = \frac{(M_1 L_w^2/2 - FL_f^3/6 + c_3 L_f + c_4)}{E_f I_f} = X_{fb}$$

$$Y_{nozzle} = \frac{(M_2 L_0^2/2 - FL_0^3/6 + c_3 L_0 + c_4)}{E_f I_f}$$

This results in a gain, from wire displacement input to nozzle displacement output, of

$$K_{wire} = \frac{Y_{nozzle}}{Y_w}$$

This gain includes the feedback gain of K_X/L_0^2 (from converting the flow to torque feedback) and the forward-loop gain of L_0/K_X (where K_X is the rotational spring rate, with units of lb/in.^{2}). The block diagram shows the static and dynamic terms for the closed-loop flow-control servovalve. The feedback wire integrates the spool velocity to position, satisfying the static requirement of spool position proportionality to input current. The flow gain of the spool completes the transformation, and the flow output is porportional to current.

The feedback wire matches the pilot stage to the boost stage, to perform either the overall pressure or the flow-control servovalve function, depending on the arrangement of the boost stage. A feedback wire is a good method of mating a pilot to a boost stage, with good overall characteristics in terms of valve static and dynamic performance. However, the use of a feedback wire introduces a disadvantage in both piece-parts-count and in interfacing the pilot with the boost stage. The nozzle flapper can be rearranged to produce a modular approach to the pilot-boost interface, if the pilot is configured as a pressure-control pilot valve.

4.3.2 Pressure-Control Pilot

Typically the boost stages associated with a feedback wire are short stroke. This implies fast response: the shorter the stroke, the more sensitive the interface becomes between the two stages. Such an interface between stages requires tight machining tolerances and entails difficult assembly procedures. The pilot stage can be altered to produce a differential pressure-control pilot.

The pilot valve, interfaced with a boost stage by the feedback wire, purposely minimizes the decentering effects of the magnetic circuit. However, it is possible to increase these decentering effects in order to cancel the mechanical stiffness of the pivot; thus the nozzle-flapper design becomes a closed-loop device, modulated by the differential pressures from the nozzle. The net stiffness K_x of the pilot was, from the previous discussion,

$$K_x = K_p - K_m + K_0 l_0^2$$

The design kept K_m to a minimum. The oil stiffness K_0 was minimized both by the small nozzles and the fact that the pivot stiffness K_p was dominant. If the decentering stiffness cancels the pivot stiffness, the net stiffness K_x becomes dependent upon the hydraulic circuit and its related stiffness. In reality there is also a null-adjustment spring (in addition to the pivot stiffness) which remains, in order to physically null the pilot stage to its center position at assembly.

The resulting stroke position of the flapper with respect to the nozzles is similar to that of the typical pilot stage. Nozzles perform a dual function when K_m cancels K_p. In addition to providing a pilot stiffness, they are the feedback mechanism for maintaining a differential pressure to the pilot-stage output. Figure 4.8 is the block diagram of the pressure-control pilot stage driving a load. This load is defined as hydraulic capacitance (C_h):

$$C_h = \frac{V}{\beta}$$

where

V = volume of oil between the pilot and load

β = bulk modulus of the oil (typically 150,000 psi)

If the load has a spring and a moving cylindrical member, the hydraulic capacitance is usually negligible compared to the equivalent capacitance of the load spring rate K_s and drive area A. The capacitance becomes, in this case,

$$C = \frac{A^2}{K_s}$$

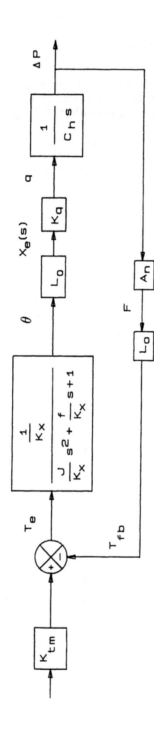

FIGURE 4.8 Differential pressure-control block diagram. Pilot is coupled with second stage (or load element) through load capacitance ($C_h = A^2/K$ for spring dominant load or $C_h = V/\beta$ for chamber volume dominant load, where A and V are the boost-stage area and volume, K is the boost-stage spring rate, and β is the bulk modulus of the oil).

152

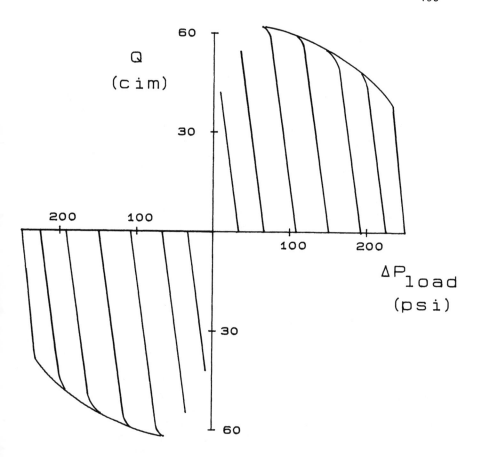

FIGURE 4.9 Pressure-control load-flow curve.

Dynamically the pilot stage is also matched to the load through
the first-order lag $1/(Ts + 1)$, where T is equal to RC and $1/R$
is the slope of a load-flow curve. Figure 4.9 is the load-flow
curve of the pressure-control pilot stage. It is desirable to have
a low hydraulic resistance (steep slopes of the load-flow curve),
since this is indicative of a component which tends to be unaffected by loading at the pilot's output. The hydraulic resistance, or output impedance, is

$$R_h = \text{output impedance} = \frac{1}{K_{pq}}$$

where K_{pq} is the slope of the load-flow curve.

For a given current level, the pilot will accommodate changes in loading by providing the appropriate flow to maintain the demanded differential pressure with very little droop. Zero droop is physically impossible. Toggling is possible if there is no net mechanical stiffness which will cause the slope of the load-flow curve to physically take on a negative value (unstable). Damping must then be introduced to ensure stability.

The capacitance is reflected in two parts of the block diagram: the dynamics through the first-order lag, and the static gain from flow to pressure. The first-order lag has time constant

$$T = R_h C = \left(\frac{1}{K_{pq}}\right)\left(\frac{A^2}{K_s}\right)$$

It is desirable to have a large bandwidth for the pilot and other stages to which it is mated. In order to accommodate this goal, the first-order lag should also have a large bandwidth (good frequency response). This is obviously set by a small time constant T. The low output impedance will definitely tend to keep the time constant T small, in comparison with other dynamic lags of the system, as long as the chosen capacitance is effectively low.

Therefore, if the spring rate of the driven stage is high, T will become small. The driven area is, however, more dominant since T is proportional to A^2. The gain, from flow output to differential pressure, is the inverse of the hydraulic or the equivalent capacitance of the load. This gain (inverse of capacitance) is forced into differential pressure by means of the feedback.

The feedback is the force created by the differential pressure on the flapper, which, because of the previously discussed magnetic circuit, allows the summing of torques as shown in Figure 4.8. Statically, if the spring rate is large or the area is small, the open-loop gain becomes large; this can be desirable for higher bandwidths. The resulting open-loop gain can, however, become too high for the dynamics of the inherent loop, resulting in a highly peaked system which must be suppressed by a component (or system) damping mechanism.

The block diagram of Figure 4.8 incorporates the boost-stage parameters of spring rate and area in the inner loop and following the closed loop. The result of the closed-loop differential pressure, driving the spring over the area of the spool, produces a spool position proportional to input current. Both types of pilots (feedback-wire and differential-pressure) drive a spool, so its stroke is also porportional to its input. The resulting flow of each style is proportional to spool stroke and therefore to input current.

Flow-control servovalves are prevalent in a wide variety of hydraulic systems. Pressure-control servovalves are not as extensively used, but they offer an approach which should not be overlooked. The pressure-control pilot stage allows a unique method of providing a two-stage valve which is modular and less expensive to manufacture. This can be carried over to pressure-control servovalves. Applications which need to control pressure (or force) are well suited for pressure-control servovalves. Pressure control can be used to effectively control inner-loop systems or other multiple-loop control schemes.

4.3.3 Pressure-Control Boost Stage

Figure 4.10 is a two-stage pressure-control servovalve. The pilot could be pressure-control or standard (which will lose proportionality, especially under load). The boost stage becomes a power amplifier in that it will reproduce the pilot pressures at a much higher flow rate than the pilot stage alone could produce. The spool and bushing could be sized also to amplify the pressures of the pilot stage. A given differential pressure input at the periphery of the spool ends will force the spool in the direction of the lower ambient pressure, say to the right. This spool position opens supply to allow output flow to the workport (C_2). Return flow from workport C_1 will exhaust to tank pressure by this same spool movement.

The pressure resulting at each workport is fed to its respective spool end. The higher pressure at the right end together with the lower pressure from the pilot will try to force the spool to the left. P_{C_2} will statically approach the supply pressure and eventually force the spool back toward its neutral position. Because of the symmetry of the design, the spool will modulate with these inputs and feedback drives; this results in a second stage which duplicates the pilot-section differential pressure at a higher flow rate (for equal areas in input and feedback).

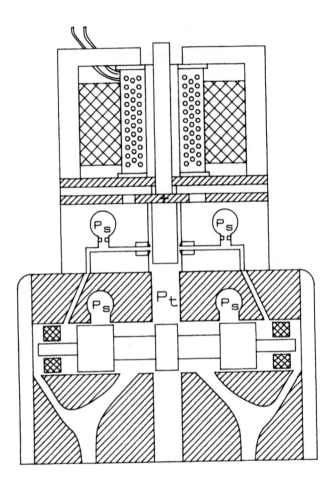

FIGURE 4.10 Two-stage pressure-control servovalve.

Another approach which provides a separate stage spool for
the feedback mechanism is shown in Figure 4.11. It eliminates
the bushing and stub diameter at the expense of the feedback
wire and feedback piston. The independent sizing of the feedback allows a potential differential pressure range not obtainable
for the single spool. Figure 4.12 is a block diagram of the valve.
    Although the pilot stage is not a differential pressure-control
pilot, the resulting servovalve is closed loop because the loop is

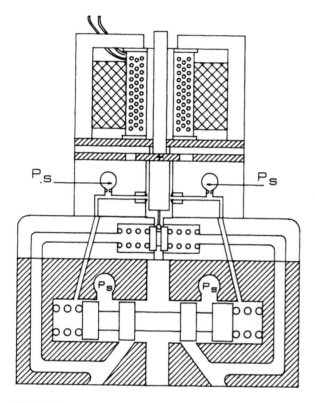

FIGURE 4.11   Alternative form of pressure-control servovalve.

closed around both the pilot and boost stages by the feedback
wire. By contrast, the pressure-control servovalve of Figure
4.10 is, in total, a differential pressure-control servovalve only
if the pilot stage also has differential pressure control (because
the two stages are in series). If the pilot stage does not have
differential pressure control, the boost stage will only amplify
the pressures of the pilot, and the pilot stage will be susceptible
to load and environmental changes.

   The pilot stage in Figure 4.11 will put out a differential pressure to the boost stage. When the spring-pressure equivalent of
the boost stage equals the pilot output, the spool will obtain a
position proportional to the differential pressure (but not necessarily proportional to input current). This spool opening,

158

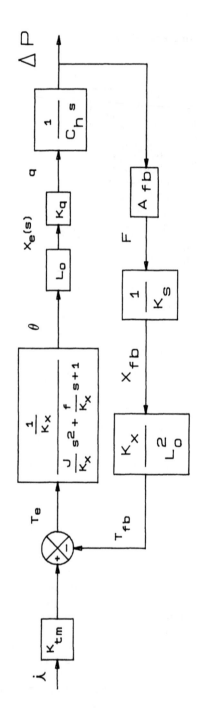

FIGURE 4.12 Block diagram of pressure-control servovalve.

say to the right, will port oil from supply out the left control
port, with return flow from the right control port exiting to the
tank. Both output ports are fed to the feedback piston. Since
the left port was opened to supply, there will be a higher pressure on the left end of the feedback piston; this moves the piston to the right.

This piston movement is reflected as piston position by the
balancing of spring forces with the differential forces created by
the feedback pressures. This rightward piston position is monitored by the feedback wire and forces a feedback torque in the
opposite direction to the initial input. The feedback wire will
balance the flapper to a position which causes the output differential pressure of the servovalve to be proportional to the input
current. The block diagram of Figure 4.12 indicates the closed-
loop nature.

A method [1] which eliminates the bushings, stub diameters,
feedback wire, and feedback spool is shown in Figure 4.13. This
approach has distinct advantages over both of the other approaches. The obvious advantage is its fewer critical components. A discussion of its operation with respect to its load-
flow curves will show its performance superiority. The key to
the benefits of Figure 4.13 is the individual spool operation for
each pilot output pressure. The resulting output is a mutual
cooperation, rather than competition of tolerances (and resulting laps) for the single spool.

Each ambient pressure forces its spool downward and opens
supply pressure to the respective output workports. This output pressure is sensed over the bottom area of the spool. This
feedback pressure will build up to the point where it will become
larger than the pilot-pressure output, and it will cause the spool
to go upward, thus cutting off supply flow and opening tank
pressure to the workport and feedback. Now the feedback pressure will eventually be lower than that of the pilot stage.

This modulation continues with each spool in a closed-loop
fashion, resulting in output pressure from the boost stage identical to their individual input pressures of the pilot stage. The
result is the same differential pressure output scale factor as
the pilot stage but with a considerable increase in flow capacity. This power amplification performance is maximized by the
individual spool modulation to preserve its input pressure.

The spool will accommodate loading commands by producing
the appropriate flow (concurrent with matching the demands
of the input pressure). This interaction allows one spool to

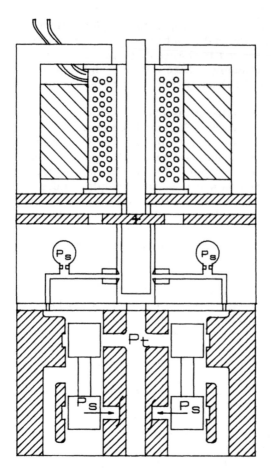

FIGURE 4.13 Two-spool pressure-control servovalve.

effectively "get out of the way" while the other spool provides
the necessary flow to the load, to match the differential pressure
commanded by the pilot. The single spool will be forced to operate with the manufactured orifices. Under some loading conditions, the fixed orifice locations will keep the valve from the ideal
performance.

Typically, pilot stages have a maximum output differential
pressure well under 1000 psi. In order to increase the differential pressure of the two-spool valve, a reduced area on the

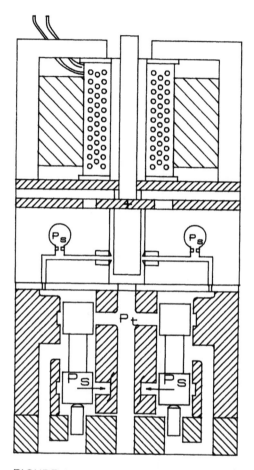

FIGURE 4.14 Two-spool pressure-amplified servovalve.

Feedback portion of the spool is necessary. Figure 4.14 shows a
two-spool pressure-control servovalve with unattached shafts at
the feedback path of each spool. The force balancing of each
spool allows a higher pressure at each output than at each input.

     The additional parts are inexpensive because the tank pres-
sure at the junction of the spool and shaft allows separate pieces,
without concentricity and with fewer fit problems in reducing
the feedback area. Figure 4.15 is the block diagram for both

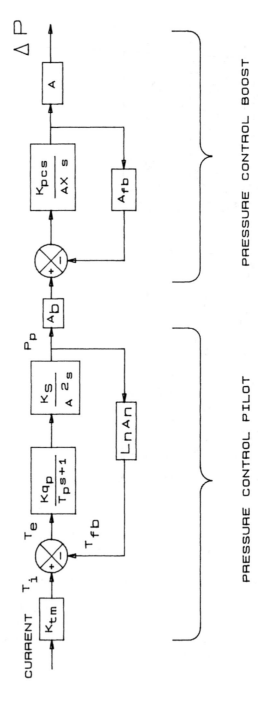

FIGURE 4.15 Block diagram of pressure-control servovalve of Figure 4.14.

versions of the two-spool pressure-control servovalves. Although each stage appears to be independent and in series, they are coupled through the capacitance of the boost stage; this results in a very responsive and stable servovalve.

The scale factor (overall static gain) of the pressure-control pilot is proportional to the torque motor gain and inversely proportional to the nozzle area. The boost-stage scale factor is the area of the spool divided by the area of the shaft. For a maximum output of 350 psi from the pilot stage and a 7:1 area ratio of the boost stage, the resulting differential pressure would be approximately 80% of supply pressure. The steep slope of the load-flow curve remains a feature of the overall valve. If the feedback area is the area of the spool, the scale factor of the second stage is unity; this results in power amplification with the identical scale factor of the pilot stage.

4.4 FLOW-CONTROL SERVOVALVES AND VALVE DESIGN CRITERIA

At first glance, valve design may appear simple. Sizing the valve for a load and to a chosen design, keeping it controllable and responsive, and maintaining stability are the guidelines to good valve design. Chapter 2 discussed sizing the valve to a load. In order to design the valve, we must know several factors about the load. Knowing only the supply pressure or only the valve pressure-drop rating is not enough; both should be known.

Flow saturation effects in flow-control servovalves and demands for steep load-flow curves in pressure-control servovalves illustrate the requirement for proper sizing of porting chambers and orifices. The loop of Figure 4.15 holds for the two-spool design wherein the scale factor becomes a function of the ratio of the boost area (A_b) to the feedback stub area (A_{fb}). The independence of spool operations allows for smooth output profiling. Although the functions of the pressure-control and flow-control servovalves are vastly different, they have similarities in design, response, and stability. The remainder of this chapter focuses on multiple-stage valving characteristics and flow-control servovalves (single- and double-spool designs).

4.4.1 Flow-Control Servovalve

Physical limitations as well as design limitations and tradeoffs determine the valve's controllability, stability and sizing. Either the electrical feedback flow-control servovalve of Figure 2.25 or the feedback-wire flow-control servovalve of Figure 4.5 can be represented by the basic dynamic block diagram in Figure 4.16, which describes minimum dynamics due to the integrator. The open-loop gain (K_0) is

$$\frac{Q}{X_p} = K_1 K_2, \quad K_0 = \frac{K_1 K_2}{A_p}, \quad K_0 = \frac{Q}{X_p A_p}$$

where

K_1 = forward loop gain; K_2 = feedback gain

X_p, A_p = piston strokes

Q = flow from the valve

The reduced form is shown in Figure 4.17. The loop portion reduces to the first-order lag, with transfer function $1/(T_s + 1)$ where $T = 1/K_0$. Dynamically, as discussed in Chapter 3, the first-order lag has an effective bandwidth (or cutoff frequency f_c) of

$$f_c = \frac{1}{2\pi T} = \frac{K_0}{2\pi} = \text{bandwidth} = 0.16 \frac{Q}{X_p A_p}$$

For a given piston area and maximum stroke, the cutoff frequency (or bandwidth) of the system is determined by the maximum flow capability of the pilot stage. Thus the valve has a velocity limit. The valve stroke also has an obvious limit which will dictate a maximum flow. The acceleration limit of the servovalve is determined by examining the sinusoidal signals

$$x = X \sin(\omega t), \quad \dot{x} = X\omega \cos(\omega t)$$

where

x = piston stroke

X = maximum stroke

$\dot{x} = v$ = velocity

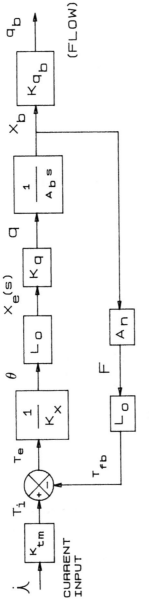

FIGURE 4.16 Minimum dynamics of flow-control servovalve.

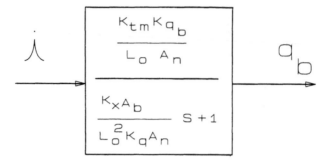

FIGURE 4.17 First-order equivalent of Figure 4.16.

$$v = 2\pi f X$$

$$f = \frac{v}{2\pi x} = \frac{Q/A}{2\pi x}$$

$$f_c = \frac{0.16 Q_{max}}{A_p X_p} = \text{velocity limit}$$

The amplitude, in a Bode plot, rolls off at f_c (the cutoff frequency) at 20 dB/decade. The acceleration is obtained by differentiating the velocity with respect to time:

$$\ddot{x} = -X\omega^2 \sin(\omega t)$$

$$\ddot{x}_{max} = X\omega^2 = X(2\pi f)^2 = g(\#g's)$$

where

g = gravitational constant = 386 in./s^2

$\#g's$ = the number of (g)'s to equal the acceleration

This maximum acceleration must provide the energy to move the mass load (m):

$$F = ma = m\ddot{x} = \frac{W(g's)}{g} = \Delta P_1 A$$

where

W   = weight of the load

ΔP_1 = differential pressure at the load

$$P_1 = \frac{W(g's)}{Ag} = \frac{0.1\ WX_f^2}{A}$$

Rearranging for the frequency, we obtain

$$f = \sqrt{\frac{\Delta P_1 A}{0.1 x W}}$$

This determines the acceleration limit of the piston. The Bode plot of the acceleration limit has a rolloff of 40 dB/decade, since the frequency-amplitude dependence contains a squared function. In order for a valve to provide enough energy to maintain an acceleration limit, the piston stroke x must be maintained and the flow limit must be adequate. Figure 4.18 is a Bode (frequency-response) plot of a multistage valve showing the stroke, velocity, and acceleration limits. The amplitude is the boost-stage stroke.

Obviously the stroke has physical limits which determine the maximum amplitude. For example, the maximum stroke corresponding to the cutoff point actually cannot be achieved because the pilot-stage flow limits the velocity (and therefore the stroke) of the boost-stage spool. Generally, the acceleration limit is determined by the flow available from the pilot stage. It is not desirable to have acceleration limits close to the flow limits. The acceleration limits at lower amplitudes will help filter out higher-frequency noise.

The boost stage can either be a boost-stage spool (which provides an amplification in terms of pressure or flow of the pilot stage) or a ram (which controls a load under position, velocity, acceleration, open-loop, or combination control). The pilot stage can itself be more than one stage. The concern is to provide enough energy to the final-stage spool or ram to obtain sufficient bandwidth for system needs.

Figure 4.19 shows two stages: a boost-stage spool (which could be the result of two or three stages or a single-stage pilot) and the actuator (ram). The ram itself could be another stage of the valve. In order to keep the bandwidth at a large

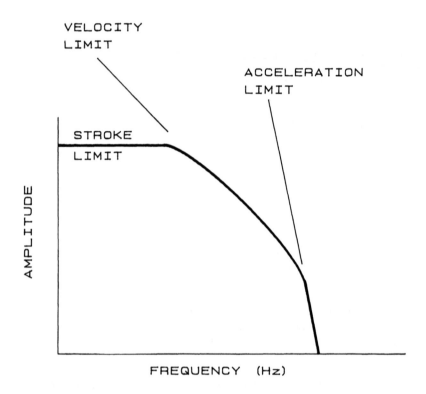

FIGURE 4.18   Frequency-response limits of flow-control servo-valve.

value, the natural frequency of the final stage must be suffi-ciently high so as not to affect the valve-ram dynamics. The natural frequency of a centered ram can be evaluated as follows. The compressibility of oil is defined by

$$K = \frac{1}{V} \frac{\Delta V}{\Delta P}$$

where the compressibility K represents a decrease in volume of a given volume when the pressure is increased. The inverse of compressibility is the bulk modulus of the oil:

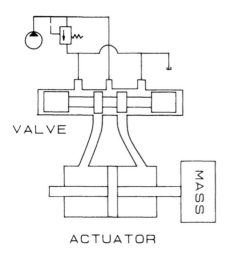

VALVE

ACTUATOR

FIGURE 4.19 Hydraulic system.

$$\beta = \frac{1}{K} = V \frac{\Delta P}{\Delta V}$$

$$\Delta V = \Delta P \frac{V}{\beta}$$

For Figure 4.19,

$$\Delta V_1 = A_p \, \Delta X_r$$

$$\Delta X_r = \frac{\Delta V_1}{A_p} = \Delta P_1 \frac{V}{\beta} \frac{1}{A_p}$$

The differential stroke, with the required differential force, determines the effective spring rate of the oil. Since the force is

$$\Delta F = A_p \Delta P = A_p (\Delta P_2 - \Delta P_1)$$

the resulting oil spring rate is given by

$$K_{oil} = \frac{\Delta F}{\Delta X} = \frac{A_p(\Delta P_2 - \Delta P_1)}{(\Delta P_1 V_1/\beta)(1/A_p)} = \frac{A_p^2\beta\Delta P}{V_1\Delta P_1} = \frac{A_p^2\beta(2\Delta P_1)}{V_1\Delta P_1}$$

$$= \frac{2A_p^2\beta}{V1/2} = \frac{4A_p^2\beta}{V} \quad \text{where } V = V_1 + V_2$$

The resulting natural frequency of the (oil-flow driven) ram is

$$F_n = \frac{\sqrt{K_{oil}/M}}{2\pi} = \frac{\sqrt{K_{oil}g/W}}{2\pi}$$

$$= \frac{\sqrt{4A_p^2\beta g/VW}}{2\pi}$$

$$= \frac{2800 A_p}{\sqrt{WV}}, \quad \text{where } g = 386 \text{ in./s, } \beta = 150,000 \text{ psi}$$

Therefore, whether the ram is a valve-stage spool or an actual ram, it is desirable to have a large piston area in order to keep its natural frequency high. The flow-velocity cutoff frequency, however, is higher with smaller piston areas. The proper balance must be maintained for the overall design. As stated previously, the pressure is the result of resistance to flow. Pressures P_1 and P_2 result from the interaction of the flow input to the load and actuator sizing. The compressibility of the oil reflects the ram's resistance to flow. The compressible flow is

$$Q_c = \frac{dV}{dt}$$

Incorporating the bulk modulus gives

$$dV = \frac{V}{\beta} dP$$

The compressible flow then becomes

$$Q_c = \frac{V}{\beta} \frac{dP}{dt}$$

The pressure buildup is therefore

$$P = \frac{\beta}{V} \int Q_c \, dt$$

The compressible flow is

$$Q_c = Q_v - V_p A_p = Q_v - \dot{X}_p A_p$$

which implies that the velocity of the ram will compress the oil from the pilot stage(s). The resulting pressures P_1 and P_2 arise from this compression, according to

$$P_1 = \frac{\beta}{V_1} \int (Q_1 - \dot{X}_p A_p)\, dt, \quad P_2 = \frac{\beta}{V_2} \int (Q_2 + \dot{X}_p A_p)\, dt$$

The flows Q_1 and Q_2 are actually a function of P_1 and P_2 according to the orifice equations

$$Q_1 = Q_m \left(\frac{X_b}{X_m}\right) \sqrt{\frac{P_s - P_1}{P_s}}$$

$$Q_2 = Q_m \left(\frac{X_b}{X_m}\right) \sqrt{\frac{P_2}{P_s}}$$

for a positive X_b. The flow rates Q_1 and Q_2, for negative boost-spool strokes, are

$$Q_1 = Q_m \left(\frac{X_b}{X_m}\right) \sqrt{\frac{P_1}{P_s}}$$

$$Q_2 = Q_m \left(\frac{X_b}{X_m}\right) \sqrt{\frac{P_s - P_2}{P_s}}$$

The load therefore will limit the velocity of the pistons; this sets the pressures and pilot output flow.

This interplay between pilot output flow and resulting pressure stresses the need for good valve design. From this discussion, the interdependence between pressure and flow must be optimized. In order to have responsive movement at the ram for a change in boost-spool stroke, without spongy transients,

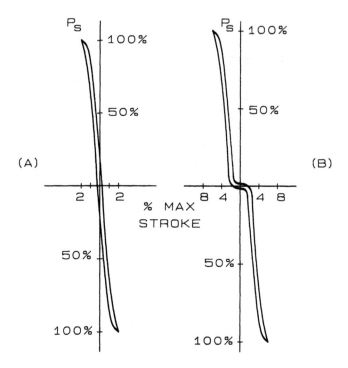

FIGURE 4.20 Pressure rise curves of flow-control servovalve.
Curve A represents line-to-line porting conditions. Curve B
result from an overlap on either pressure or tank porting edges.

the boost spool must have a good pressure rise and maintain ideal
slopes in its output load-flow curves.

The pressure rise is tested at blocked output ports (the ram
essentially is fixed in position) with leakage at only the boost
spool itself. Figure 4.20 represents typical pressure-rise curves.
A good valve design will develop full system pressure in less than
2% of the maximum rated stroke. Since there is no moving output,
the pressures P_1 and P_2 result from the capacitance of the oil:

$$P_1 = \frac{\beta}{V_1} \int Q_1 \, dt, \quad P_2 = \frac{\beta}{V_2} \int Q_2 \, dt$$

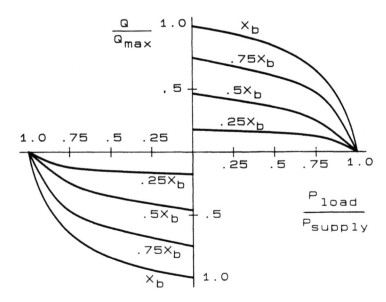

FIGURE 4.21 Load-flow curves of flow-control servovalve.

The flows Q_1 and Q_2 are both zero output in this case. Leakage flow from P_s to tank exists as a result of the clearances, laps, sharp-edge orificing, and port shaping between the spool and its bores. Curve A is representative of line-to-line spool-to-bore-porting dimensions, with diametrical clearances of 0.0005 in. per inch of diameter. Curve B results from changing the lap between supply and output by approximately 5% of rated stroke, with an underlap (opened orifice) between the output port and tank pressure equal to the supply overlap. Section 4.4.4 discusses a two-spool flow-control servovalve which can be adjusted to obtain either A or B pressure-rise curves.

Curves A and B are necessary to create a stiff link between the valve and its output stage. The load-flow curves (obtained by plotting the flow equations for Q_1 and Q_2) are shown in Figure 4.21. Various values of X_b are shown on the same plot for both directions of flow. The ideal plots would have horizontal slopes, indicating that the valve is unaffected by attaching it to a load. In regard to spool design, saturation effects, and physical limits, the curves tend to be horizontal at low load pressures and converge to zero flow when the load pressure equals the supply pressure.

Tradeoffs from ideal must be considered in order to maintain stability in valve design. Linearization of the orifice flow equation will indicate the need to create a balance in valve design with regard to valve stiffness. The flow equation simplifies to

$$Q = f(P_s, P_c, X) = K_q X_b - K_{pq} \Delta P_1 \quad \text{where } \Delta P_1 \text{ is the load pressure}$$

The flow has also been shown to be equal to the ram velocity flow plus the compressible flow:

$$Q = A_p \dot{X}_p + \frac{V}{4\beta} \frac{d(\Delta P_1)}{dt}$$

For a load which has effectively zero friction and which is dominated by the mass attached to the ram, the load differential pressure becomes

$$\Delta P_1 = \frac{M \ddot{X}_p}{A_p}$$

Equating the flow equations, we obtain

$$A_p \dot{X}_p + \frac{V}{4\beta} \frac{d(\Delta P_1)}{dt} = K_p X_b - K_{pq} \Delta P_1$$

Substituting the mass load for the differential pressure gives

$$A_p \dot{X}_p + \frac{V}{4\beta} \frac{M}{A_p} \dddot{X}_p = K_q X_b - \frac{K_{pq} M}{A_p} \ddot{X}_p$$

In Laplace form, this becomes

$$A_p s X_p(s) + \frac{V}{4\beta} \frac{M}{A_p} s^3 X_p(s) = K_q X_b(s) - \frac{K_{pq} M}{A_p} s^2 X_p(s)$$

Rearranging in terms of input $X_b(s)$ and output $X_p(s)$ yields

$$X_p(s)\left\{\frac{V}{4\beta}\frac{M}{A_p^2}s^3 + \frac{K_{pq}M}{A_p^2}s^2 + s\right\} = \frac{K_q X_b(s)}{A_p}$$

or

$$\frac{X_p(s)}{X_b(s)} = \frac{K_q/A_p}{s\left[(V/4\beta\ M/A_p^2)s^2 + (K_{pq}M/A_p^2)s + 1\right]}$$

Figure 4.22 is the block diagram of the spool driving the ram. The open loop would result in ram velocity. It is assumed that a feedback of mechanical linkage, feedback wire, or electrical transducer forces the output to ram position. The feedback is represented as K_{fb}. The valve and ram variables which are critical to the design, and which can be controlled, are combined in one term, K_c, the critical valve gain:

$$K_c = \frac{A_p K_q}{K_{pq}}$$

Substituting this gain into the preceding equation results in the transfer function between the boost-spool position and piston position of the ram:

$$\frac{X_p(s)}{X_b(s)} = \frac{K_q/A_p}{s\left[M/K_{oil}s^2 + (K_q/A_p)M/K_c s + 1\right]} \quad \text{where } K_{oil} = \frac{4\beta A_p^2}{V}$$

The natural frequency and damping ratios equate to

$$\omega_n^2 = \frac{K_{oil}}{M} \qquad \frac{2\zeta}{\omega_n} = \left(\frac{K_q}{A_p}\right)\frac{M}{K_c}$$

$$\omega_n = \sqrt{\frac{K_{oil}}{M}} \qquad \zeta = \frac{\omega_n}{2}\left(\frac{K_q}{A_p}\right)\frac{M}{K_c}$$

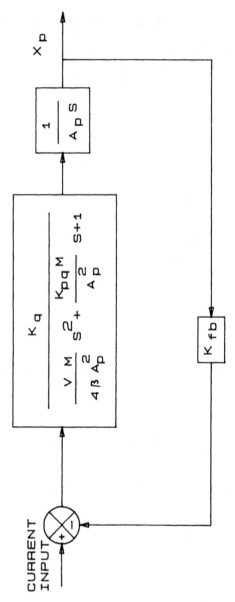

FIGURE 4.22 Position-control loop.

$$= \sqrt{\frac{K_{oil}}{M}} \left(\frac{K_q}{A_p}\right) \frac{M}{2K_c}$$

$$= \sqrt{\frac{\beta M}{V}} \frac{K_{pq}}{A_p}$$

The transfer function for the closed loop between the valve and ram is

$$\text{T.F.} = \frac{X_p(s)}{X_b(s)} = \frac{K_q/A_p}{\left(M/K_{oil}\right)s^3 + \left(K_q/A_p\right)\left(M/K_c\right)s^2 + s + K_{fb}K_q/A_p}$$

With this transfer function, we can establish a stability guideline by applying Routh's stability criterion to the transfer function:

$$s^3 \qquad \frac{M}{K_{oil}} \qquad 1$$

$$s^2 \qquad \left(\frac{K_q}{A_p}\right)\frac{M}{K_c} \qquad K_{fb}\left(\frac{K_q}{A_p}\right)$$

$$s^1 \qquad b_1$$

$$s^0 \qquad c_1$$

The constants b_1 and c_1 are defined by

$$b_1 = \frac{\left(K_q/A_p\right)M/K_c - M/K_{oil}\left\{K_{fb}\left(K_q/A_p\right)\right\}}{K_q/A_p \; M/K_c}$$

$$c_1 = K_{fb}\frac{K_q}{A_p}$$

As discussed in Chapter 3, the coefficients of the first column must be greater than zero to avoid instability. Therefore

$$\left(\frac{K_q}{A_p}\right)\frac{M}{K_c} > \frac{M}{K_{oil}}\left\{K_{fb}K_q A_p\right\}$$

$$\frac{1}{K_c} > \frac{K_{fb}}{K_{oil}} \quad \text{or} \quad K_c < \frac{K_{oil}}{K_{fb}} \quad \text{or} \quad \frac{A_p K_q}{K_{pq}} < \frac{4\beta A_p}{V K_{fb}}$$

Thus the valve parameters are set by (similar to [3])

$$\frac{K_q}{K_{pq}} < \frac{4\beta A_p}{V K_{fb}}$$

4.4.2 Flow Forces and Damping Needs

If the feedback linkage has a one-to-one stroke relationship, K_c must be less than K_{oil} for a stable valve-ram system. The ram stroke is typically much larger than the spool stroke, resulting in a small value of K_{fb}. It is desirable to keep the critical gain K_c and K_{oil} high.

The damping ratio ζ becomes small for a large K_c; this will amplify the stability effects of this inequality's demands on the design. Since $K_c = A_p K_q / K_{pq}$ (with A_p set by load requirements), only the valve flow gain K_q and the load-flow slopes can be controlled by design to maintain a responsive, stable system.

Other means of damping can add to the stability solution. The flow gain can add valve damping by port shaping. Longer strokes will add stability but hamper response. The load-flow curve can obtain steeper slopes by minimizing spool flow forces (to be discussed later) by spool shaping or by dwarfing them with large drive forces. The problem is amplified for short ram strokes and small ram areas. Long hose lengths from valve to ram will increase the oil volume and therefore decrease K_{oil} (the oil stiffness); this makes the stability problem worse.

Rather than downgrading a valve by introducing friction or by reducing the effectiveness of the flow gain (K_q), one should try to attack the source of the problem. The responsive, tight-tolerance valves lack damping. The low slopes of the load-flow curves, especially at lower load pressures and spool strokes, are worsened by flow forces. If the flow forces and damping are addressed, the results will be fruitful.

Flow forces can be minimized by spool contouring, orifice port shaping, or covering them with excessive drive forces or large spring rates. The damping can be accomplished by orifice port shaping, frictional loading, and hydraulic circuitry. On a system level, damping can be introduced electrically and electromagnetically at the pilot stage. Flow forces exist in all hydraulic valving situations from nozzle-flapper stages to relief valves and high-flow spool valves. Their consequences, if untreated, are oscillations and unstable performance.

One porting edge of the boost-stage spool(s) is shown in Figure 4.23. From Bernouli's equation, the pressure near the orifice is smaller than in other chambers within the spool and for its exit area, because the velocity is higher at the orifice. This appears to be double talk! First, flow necessitates a differential pressure drop with a lowering of pressure in the direction of flow. However, with the lowest pressure at the orifice (especially at small strokes), the flow would appear to be reversed. With absolute chamber pressures before and after the orifice indicating a flow in the direction of the intended flow of the orifice, there exists a potential stability problem.

The lower chamber pressure at the region of the orifice tends to create a leftward force on the spool; this tends to close off the spool from delivering flow. This spool closure opposes the

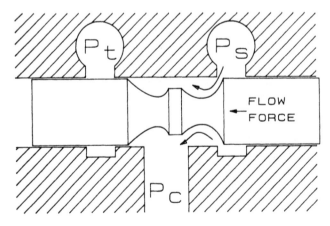

FIGURE 4.23 Flow-force-compensated spool shaping. Flow from supply to control port.

command to open the orifice by the pilot-stage differential pressure. The result is chatter between the pilot closed-loop commands and the orifice operation. This chatter, amplified through the hydraulic valving and load, will become very excessive and unstable.

To minimize the effect of the flow force F_f, either a balancing force or equalizing pressure effect (of the orifice within the spool chamber) must be pursued.

The balancing force could exist as a large spring rate to effectively "cover" the effects of the flow force. If the valve were free of springs, the input pressure must be high enough, or the area of the spool must be large enough, to counteract the flow forces. The most common method is to shape the spool to change the velocity and flow profiles to minimize the flow forces. Figures 4.23 and 4.24 represent a spool which is contoured.

Proportional valves typically have been mechanically activated, necessitating low flow forces to obtain the best "feel" for a human interface with the hydraulic components. There has been extensive research in spool shaping and cutting while maintaining low-cost production. The classical means of spool shaping was carried on by Von Mises, who determined that the flow forces were approximated by the fact that the mass flow rate is constant.

$$F_f = KA_0 \Delta P \cos(\phi)$$

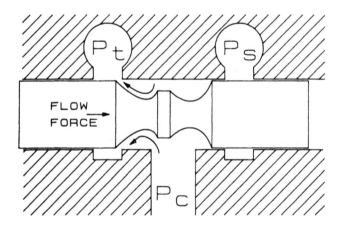

FIGURE 4.24 Return flow spool-contoured flow compensation.

where

K = constant from the orifice flow equation $Q = KA_0\sqrt{\Delta P}$

ΔP = pressure drop across the orifice = $P_s - P_c$

A_0 = orifice area = $\pi DX\beta$

β = portion of spool (or body) periphery used for porting

X = spool stroke

ϕ = angle of discharge which the fluid takes at the orifice

An estimate of these flow forces, from [4], is

$F = 0.43A_0\Delta P_v$ = net flow force (lb)

where P_v is the pressure drop across the orifice and A_0 is the orifice area.

From the investigation of the load-flow curves, low strokes, and low load pressures at the region of flow force, and the fact that there are clearances between the spool and bore, the flow-force equation should be used as a figure of merit. Experimentation is the means of determining actual forces. Through the aspects of experimentation, force estimation, spool shaping (as implied in Figure 4.24), sizing, and critical gain K_c, the solution to minimizing flow forces can be resolved.

Damping is required in all systems. As indicated, damping can aid the flow-force problem as well as other stability problems associated with hydraulic valving. It has been shown [5] that the damping coefficient (C_p) obtained by the damping chamber (of Figure 4.25) is

$$C_p = \frac{8\pi\mu L}{A_c}\left(\frac{A_p}{A_c}\right)^2$$

where

A_p = piston area

A_c = damping chamber area

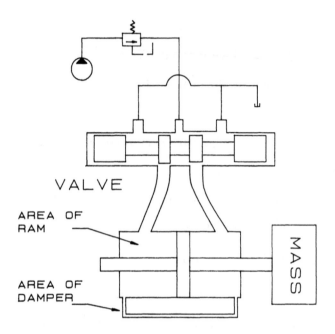

FIGURE 4.25 Hydraulic damping in valve-actuator system.

This equation holds as long as the following conditions are met:

$$\frac{\nu d \rho}{\mu} < 2000$$

where

   d = damping chamber diameter

   ρ = density

   μ = dynamic viscosity

$$\sqrt{\frac{R^2 \rho \omega}{2\mu}} < 2$$

where

   R = damping chamber radius

   ω = vibration frequency (rad/s)

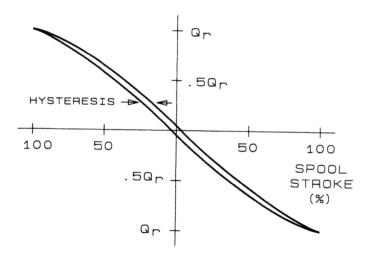

FIGURE 4.26 Flow-control servovalve flow gain. Q_r is the rated flow.

This damping technique can be employed at three locations in the two-spool pressure-control servovalve (see Figure 4.13). It can be placed between the two input ports. This will be statically handled by the closed-loop operation of the pilot. With no input signal to the pilot, each ambient pressure is the same; therefore, no bleed problems are created. The same scheme can be employed at the output ports of the pressure-control servovalve, since the total valve is also a closed-loop system. Since each output port duplicates its input port, a damping chamber can be placed between the input and output of each spool.

4.4.3 Hydrostatic Bearings

Friction is normally present at the spool, and it can increase the damping and remove oscillation problems associated with flow forces. It can be increased at the expense of large hysteresis in the flow-gain curve of the flow-control servovalve. This excessive hysteresis is typically not acceptable in servovalves, so efforts are made to decrease it. The typical flow gain of a servovalve is shown in Figure 4.26. The better the bearing is between the spool and bore, the better the hysteresis of the flow gain.

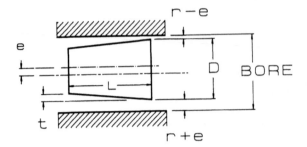

FIGURE 4.27 Hydrostatic tapered bearing.

The tapered bearing of Figure 4.27 is the answer. The centering flow forces as plotted in Figure 4.28 can be shown [6] to be

$$F_c = \frac{K_1 t/c}{e/c} \left\{ \frac{2 + t/c}{\sqrt{(2 + t/c)^2 - 4(e/c)^2}} - 1 \right\} - \frac{K_2(1/tc)}{e/c}$$

$$\left[\frac{\sqrt{(1 + t/c)^2 - (e/c)^2} - \sqrt{1 - (e/c)^2}}{t/c} - \frac{2 + t/c}{\sqrt{(2 + t/c)^2 - 4(e/c)^2}} \right]$$

where

$K_1 = \pi DL\Delta P$

$K_2 = 6\pi \mu DVL^2$

t = radial taper over the length (L) of the bearing

c = radial clearance between spool and bore

e = eccentricity of the spool centerline with respect to the bore

V = velocity of the spool

With the tapered bearing comes an increase in spool leakage, shown in Figure 4.29, calculated from

$$Q_1 = K_3 \left[\frac{2(1 + t/c)^2}{2 + t/c} + \left(2 + \frac{t}{c}\right)\frac{3}{4}\left(\frac{e}{c}\right)^2 + K_4 \left\{ \frac{2 + t/c}{\sqrt{(2 + t/c)^2 - 4(e/c)^2}} - 1 \right\} \right]$$

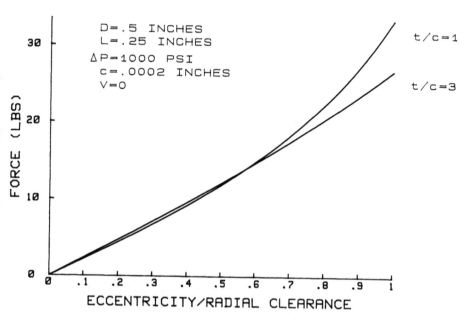

FIGURE 4.28 Tapered-bearing centering forces as a function of spool offset within bore.

where

$$K_3 = \frac{\pi Dc^3 (P_2 - P_1)}{12\mu L}$$

$$K_4 = \frac{(t/c)^4}{8(2 + t/c)}$$

The flow gain of Figure 4.26 shows a neutral stroke region which can be changed to include a dead band, or region of no flow for a spool movement. Often the profile at neutral has little to no dead band. Providing optimal laps becomes important for applications which vary from open loop to closed loop.

4.4.4 Two-Spool Flow-Control Servovalve

Lap adjustability is possible, at a cost savings, with the two-spool valve design [7,8] shown in Figure 4.30. The boost stage is driven

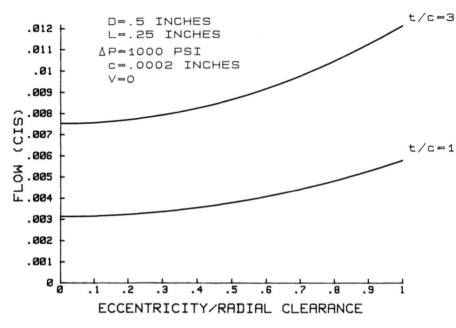

FIGURE 4.29 Tapered-bearing leakage.

by the same pilot stage as used with the single-spool flow-control servovalve of Figure 4.5. The pilot stage has differential pressure output which is continually monitored and dynamically maintained by the closed-loop feature of the magnetic and hydraulic circuits. Utilizing the power of the differential pressure-control pilot stage, both spools of the boost stage function together to use one porting edge each for metering flow in and out of the valve.

For $P_1 > P_2$, both spools go downward because of the pressure imbalance. The spools will stop at a position set by the spring (which balances the differential pressure) proportional to the electrical input. This downward movement opens supply pressure (P_S) to the left spool, allows flow (Q_a) to the workport and back into the right spool, and exhausts flow through the orifice to the tank pressure (P_t). Reversal of the electrical input signal makes $P_2 > P_1$; this results in upward spool movement and a reversal of flow through the valve.

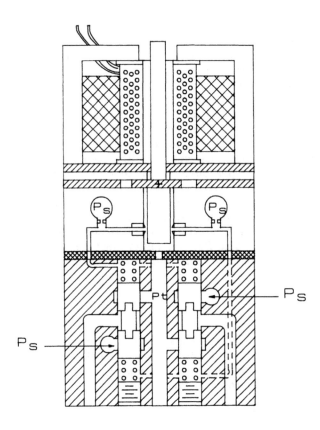

FIGURE 4.30 Two-spool flow-control servovalve.

The main advantage of the two-spool style is the independence of porting edges as compared to the conventional single-spool valve. The conventional valve of Figure 4.5 has three critical dimensions all interrelated to each other (plus three more with respect to their bore locations). The two-spool configuration has only one critical edge per spool and bore. This important difference is advantageous for both the fabrication of the critical edges and matching of the porting edges of the spool with the bore at assembly.

Flexibility is built into the two-spool valve because of the individual null adjustments. Each spool can be set to obtain a

desired lap condition at the supply and tank port edges. The
null pressures (resulting from no electrical input signal) are
therefore set by the resultant of these lap conditions. The pressure rise curve and flow profile can easily be adjusted to fit any
application. The null pressure of the conventional valve is set
by the fabricated laps and is not adjustable.

In addition to the two-spool cost savings and adjustable laps
and neutral output pressures, the two-spool valve has a fail-safe
feature if one spool becomes stuck due to contamination. If the
input is then commanded to produce zero flow, the pilot stage
will command zero differential pressure. Even though one spool
may be stuck in position, the other spool will return to neutral.
Since the four-way valve requires two orifices in series (one in
each spool), there will be no output flow with only one spool at
neutral; this creates a fail-safe mechanism which is not possible
with the single-spool valve. The flow capability must be matched
to the sizing of the spools and the body and to the porting configuration.

4.4.5 Flow Sizing

Spool valves, whether single-stage or boost-stage, pressure- or
flow-control configuration, must be sized to meet the flow requirements of the load and the flow supplied by the pump. The orifice
equation sets the initial requirements for flow:

$$Q = C_d A_0 \sqrt{\frac{2\Delta p}{\rho}} = kA_0 \sqrt{\Delta P_v}, \qquad A_0 = \pi D X \beta$$

where

$K \quad = C_d \sqrt{2/\rho}$

C_d = discharge coefficient = 0.67 or sharp-edge orifice

ρ = oil density

ΔP = differential pressure across the orifice(s)

A_0 = orifice area

D = spool diameter

β = portion of the spool's periphery used in orificing ($0 < \beta \leq 1$)

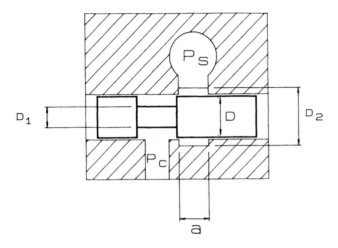

FIGURE 4.31 Spool-flow sizing.

Obviously other restrictions to flow must be greater than the maximum rated orifice in order for the orifice to be effective. If not, that restriction would control the flow. The main restrictions to flow of a servovalve are the chamber cross-sectional area of the supply, tank, and control ports. Figure 4.31 represents a spool, or portion of a spool, which orifices a high-pressure source (P_S) to a lower-pressure output (P_C). The annular area between the bore and the spool shaft and the cross-sectional area undercut in the bore at P_S must provide an adequate passageway for the oil. The passages should have at least twice the maximum orifice area, with quadruple sizing more desirable.

The flow porting within the body at supply, tank, and control ports lends to larger porting, whereas the spool sizing tends to dictate smaller areas. As a rule of thumb, the shaft diameter D_1 should be at least half the spool diameter (D). For flow saturation effects, the shaft should usually be as small as possible. Following this rule of thumb, the spool diameter becomes $D = 2D_1$. With an area ratio of twice the maximum orifice area, the cross-sectional area between the bore and shaft is

$$A_s = \frac{(D^2 - D_1^2)\pi}{4}$$

The flow path around the spool at P_s should provide from two to four times the maximum orifice area. This sizing ends up in a tradeoff in axial length (a) of the spool and body length versus stress and deflection of the groove (with respect to other grooves and ports within the body). If the upper portion were sized to an equivalent area of the maximum orifice, the total cross-sectional area would be twice the maximum orifice area. Additional area can be accomplished by grooving the spool.

The overall saturation effects can be adequately estimated by treating the orifice and passageways as a series of orifices. The equivalent orifice for simulating saturation for the typical four-way valve is

$$A_0(eq) = \left(\frac{2}{A_p^2} + \frac{2}{A_c^2} + \frac{2}{A_r^2} + \frac{2}{A_s^2} + \frac{2}{A_0^2} \right)^{-1/2}$$

where

$A_0(eq)$ = equivalent orifice of the orifices in series

A_p = supply pressure port (and return port)

A_c = equivalent area of the control port passage

A_r = (restrictive) area around the spool

A_s = cross-sectional area between the bore and spool shaft

A_0 = orifice area from spool stroking

The orifice equation relates the flow output to the differential pressure. For a four-way servovalve (two spool-stroking orifices), the orifice equation for valve flow is

$$Q = KA_0\sqrt{\Delta P_v}, \qquad A_0 = \pi DX\beta$$

where

K $= C_d\sqrt{1/\rho}$

ΔP_v = pressure drop across the two metering edges

4.5 CONCLUSION

Valving may be the key control element because it provides the transformation of electrical input to control output pressure and flow originating from the pump. Valves range from slow-acting, large dead-band, proportional valves to load-sensing valving functions and high-response systems. Common to all of their designs and requirements, the valve converts power into a form used by the needs of the system. It is effectively only an element of the plant to be controlled.

Assuming it is adequately designed for a system, the system itself must contain variables which control and improve the overall response within the stability requirements. Often, the combination of the valve with other elements is not enough for the system desired. Compensation is necessary to maintain stability while maintaining adequate response of the system.

BIBLIOGRAPHY

1. Merritt, Herbert E. *Hydraulic Control Systems*, Wiley, 1967.

2. Anderson, Wayne R. Pressure and flow control servovalves – a unique approach, in *Proceedings of the 40th Conference on Fluid Power*, Vol. 38, pp. 199–206.

3. McCloy, D., and Martin, H. R. The Control of Fluid Power, Halsted Press (a Division of John Wiley & Sons, Inc.), New York, N.Y., 1973, pp. 116–130.

4. Blackburn, John F. *Fluid Power Control*, Technology Press of M.I.T. and Wiley, 1960.

5. Harris, Cyril M., and Crede, Charles E. Flow of viscous compressible fluid through conduit, in *Shock and Vibration Handbook*, section 32, 1967, pp. 30–32.

6. Viersma, T. J. Frictionless Hydraulic Motors, Technical College in Delft, The Netherlands.

7. Yeaple, Frank. Twin spool knocks cost out of servovalve, *Design News*, vol. 40 (1984), 118–119.

8. Beercheck, Richard C. Electrohydraulic valves: key links between electronic brains and hydraulic brawn, *Machine Design*, Vol. 57 (1985), 56–58.

5

Controlling Stability

5.1 OPEN-LOOP INVESTIGATION

Stability is the major concern of any type of control system. A
system is stable if, after a disturbance to the system occurs, the
system eventually returns to its equilibrium state. The equilibrium state may reflect a null spool position in a servovalve, an
output pressure corresponding to a set input command, a motor

speed proportional to a handle position or to other conditions set
by the requirements of the system.

Electrical controllers combined with electrohydraulics have
various advantages over hydromechanical controllers in establishing stability. Compensating networks exist in hydraulic
systems. In either case, the general block diagram of a closed-
loop system and its major elements can be represented by Figure 5.1.

For the most part, we have described the plant G(s) or system to be controlled. This has been either a component, such
as a valve, or components tied together to form a system. The
feedback elements control the output C(s) by monitoring it and
informing the controller of its state. The controller uses the
information from its feedback (error signal) to provide the appropriate correction signal to the plant via the controlling elements.

The controller, whether analog or digital, electrical or hydromechanical, must provide the proper signal to optimize output
response consistent with stability requirements. The controlling elements are, in a sense, a part of the plant. An electrohydraulic multistage servovalve may be added to a system G(s) to
enhance the plant's response and tie the plant to an electrical
signal. In this case, the electrohydraulic servovalve is added
as a "controlling element."

In other systems, a servovalve may exist as part of the plant.
The controlling element adds to the complexity of the plant, and
efforts should be made to keep its dynamic effects small in comparison to the plant. The controller will be used to compensate
the dynamic losses of the plant.

The general system block diagram is shown in Figure 5.2. The
disturbance is actually another input to the system. The closed-
loop transfer function for the input R(s) was shown to be

$$\frac{C(s)}{R(s)} = \frac{G(s)}{1 + G(s)H(s)}$$

For the input L(s), the transfer function for the disturbance is
derived assuming the input R(s) is not present. The resulting
transfer function for the disturbance is

$$\frac{C(s)}{L(s)} = \frac{G_p(s)}{1 + G(s)H(s)}$$

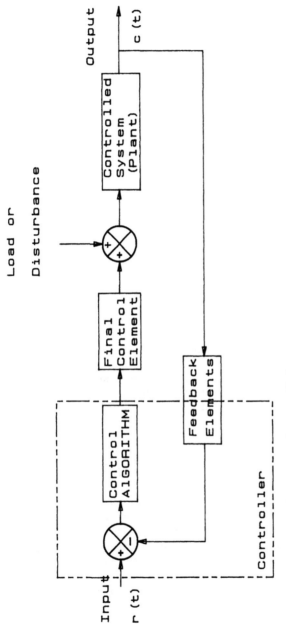

FIGURE 5.1 Basic elements of controlled system.

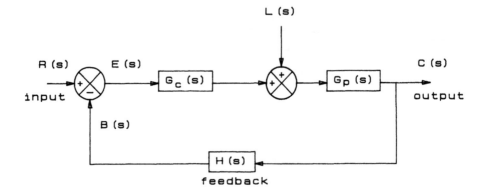

FIGURE 5.2 s-domain notation for Figure 5.1.

where

$$G(s) \ = G_c(s)G_p(s)$$
$$G_c(s) = \text{controller element(s)}$$
$$G_p(s) = \text{plant or items to be controlled}$$

A linear system allows superposition in that the output C(s) is the combination due to both reference and disturbance inputs. The result of including both inputs is the "operational equation"

$$C(s) = \frac{G(s)}{1 + G(s)H(s)} R(s) + \frac{G_p(s)}{1 + G(s)H(s)} L(s)$$

The key to control systems is the controlling element, since it is here that the supply of energy is controlled. For a system which has a servovalve as its controlling element, the energy which must be controlled is the high-pressure oil supply pressure source. For a low-level input, the valve precisely controls a high-power-level output.

Matching the valve to the plant is critical for good loop response and stability. If the matching is improper, the system will become loose or ineffective when working against loads or disturbances on the plant. This matching is expressed as impedance matching. The pressure-control pilot stage discussed in Chapter 4 can be mated to a variety of boost stages. The interface between the stages, even though the physical fit is modular and closed loop, is dependent on the impedance match between the pilot and boost stages.

Ideally, the controlling element and every other element in a block diagram should require zero energy from its input signal; that is, each should have infinite input impedance and zero output impedance. However, the impedances must be properly matched between the final controlling element and the driven element for maximum power transfer to the plant. This final controlling element must provide energy to the plant at the rate commanded by the controller.

5.2 VELOCITY FEEDBACK

Velocity feedback with a controller loop is a method of maximizing performance with respect to response and stability. The velocity output valve-actuator system without linkage feedback is desirable because it is a first-order lag (open-loop portion of Figure 3.11). However, it is open loop and subject to disturbances. When the linkage was added to close the loop to change the output from velocity to position, the order of the system was increased to a second-order lag (closed loop of Figure 3.11).

The low-damping component could create excessive peaking and instability, especially when used with other components in a more complete system. Loops which demand high response but lack damping can obtain damping in several forms, as discussed in Chapter 4. Many methods can be used to compensate and enhance system response and stability. If the velocity and position were used for control, the system would retain benefits of both.

The best method of obtaining position and velocity control is to change the system to include an electrical controller to handle the double-loop closure. If the velocity is measured with a velocity transducer and if the position is measured with a position transducer, the block diagram of Figure 5.3 would result. By block-diagram reduction, it reduces from Figure 5.4 to Figure 5.5.

Recall from the introduction that the human interface also controlled both velocity and position with limitations in response.

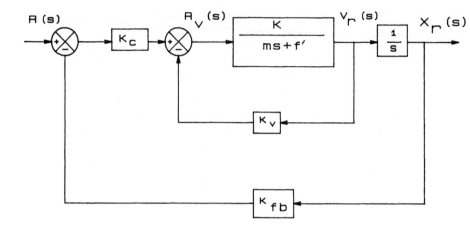

FIGURE 5.3 Velocity feedback inner-loop control.

The human feedback link in controlling these terms was adequate for low-bandwidth loops. The transfer function for the valve actuator with linkage is (refer to the end of Section 3.2).

$$\frac{C(s)}{R(s)} = \frac{K}{ms^2 + f's + KK_{\text{link}}}$$

The transfer function for Figure 5.3 is

$$\frac{C(s)}{R(s)} = \frac{K/s(ms + f')}{1 + K(K_v s + 1)/s(ms + f')} = \frac{K}{ms^2 + f's + (KK_v)s + K}$$

$$= \frac{K}{ms^2 + (f' + KK_v)s + K} = \frac{1}{(m/K)s^2 + [(f' + KK_v)/K]s + 1}$$

Comparing these two transfer functions with the general transfer function of second order,

$$\frac{C(s)}{R(s)} = \frac{1}{(s/\omega_n)^2 + (2\zeta/\omega_n)s + 1}$$

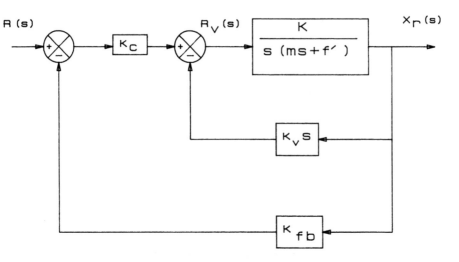

FIGURE 5.4 Reduced block diagram of Figure 5.3.

it becomes obvious that the gain of the velocity transducer K_v does not affect the natural frequency ω_n, but it contributes to the damping ratio; this produces a system which can be tuned for optimal damping (and therefore stability and response).

The inner loop becomes a valuable tool for tuning a control system. It can be used to increase the capability of a loop in terms of higher bandwidth and controllability. The feedback itself can be used to shape the dynamics of a loop.

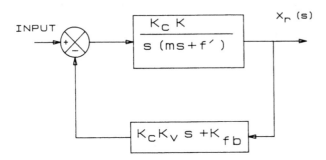

FIGURE 5.5 Lead feedback.

The controller, which is often integral with the feedback, becomes the controlling element. The controller ties together the feedback with the command to the controlling elements to drive the plant. A combination of feedback elements with forward-loop elements can "shape" a system from unstable to stable. This shaping can be accomplished by matching the open-loop response (theoretical or experimental or a combination) with the Bode plot stability requirement. Gain and phase margins dictate dynamic and static terms necessary to bring the system into a region of stability. The controller can be tuned to fit the gains and dynamic needs of the plant.

5.3 PID

A widely used technique of matching the needs of the plant is the PID (proportional plus integral plus derivative) algorithm. PID control schemes combine an integrator with derivative and proportional elements. The integrator produces zero output droop, as explained in Chapter 3. The PID combination is shown in Figure 5.6. The block diagram reduces to

$$K_p \left\{ K_x + \frac{1}{T_i s} + T_d s \right\} = K_p \frac{K_x s + 1/T_i + T_d s^2}{s}$$

$$= \frac{K_p}{T_i s} \left\{ T_i T_d s^2 + T_i K_x s + 1 \right\}$$

This is a second-order lead. K_x is typically equal to unity. These parameters are equated to the second-order natural frequency and damping ratio by

$$\omega_n^2 = \frac{1}{T_i T_d} \quad \text{and} \quad T_i K_x = \frac{2\zeta}{\omega_n}$$

$$F_n = \frac{1}{2\pi / \sqrt{T_i T_d}}, \quad \zeta = \frac{T_i K_x \omega_n}{2} = \frac{T_i K_x}{2\sqrt{T_i T_d}} = \frac{K_x}{2\sqrt{T_i T_d}}$$

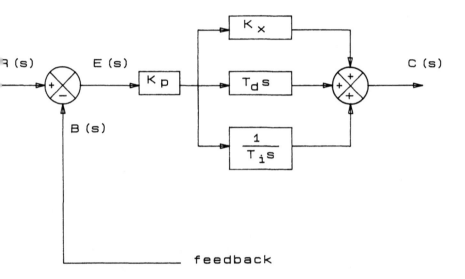

FIGURE 5.6 PID controller.

It is generally difficult to adjust the parameters K_p, T_i, and T_d to a system. If T_i is altered, it affects the static gain, the natural frequency of the second-order lead, and its damping ratio. T_d will affect the damping ratio and the natural frequency. If T_i is set to a value, K_p can be adjusted to become the static gain of the controller. With T_i set, T_d will dictate the natural frequency of the second-order lead.

The damping ratio is, however, a function of T_d. It would be desirable to adjust the damping ratio separately. K_x, if present, can fulfill this requirement as long as its value matches $\sqrt{T_d}$. If K_x matches $\sqrt{T_d}$ as T_d is changed to fit the second-order lead to the system, K_x can be changed beyond the match to control the damping ratio independently. In an analog electrical system, this can be difficult. In a digital system, the desired damping ratio can be entered, with the microprocessor matching K_x to the changes in T_d, as

$$K_x = 2\zeta \sqrt{\frac{T_d}{T_i}}$$

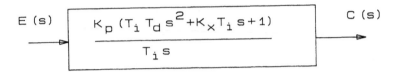

FIGURE 5.7 Dynamic equivalent PID of Figure 5.6.

The PID reduces to a typical second-order lead with integration as shown in Figure 5.7. If the derivative time constant were zero, the controller would reduce to the first-order lead transfer function $K_p(T_i K_x s + 1)/T_i s$. T_i once again affects the static and dynamic terms; this makes T_i difficult to adjust. If T_i is set to a value and not changed, the static gain is set by K_p and the dynamic lead is adjusted by K_x. If the integrator time constant, instead of the derivative time constant, is zero, the block diagram reduces to

$$K_p(K_x + T_d s) = \frac{K_p}{K_x}\left(\frac{T_d}{K_x}s + 1\right)$$

A steady-state closed-loop droop due to the absence of an integrator will then occur. Some systems will have an integrator, caused by the feedback used. Such a system still retains the benefits of this controller because of the dynamic first-order lead. In this situation, K_x should be set to unity. Figure 5.8 is the PID within a closed loop.

5.4 PSEUDO-INTEGRATOR

If high bandwidth is not the overriding concern for a system, a first-order lag can be utilized as a "pseudo-integrator" with compensation benefits. Figure 5.9 is a block-diagram representation of a servovalve driving a hydraulic motor in a closed-loop mode; it produces motor speed proportional to a voltage command. The plant is the valve (with the static flow gain K_q and its associated dynamics) and the motor (with its gain $1/D_m$, which is the volumetric displacement). The feedback element is a velocity

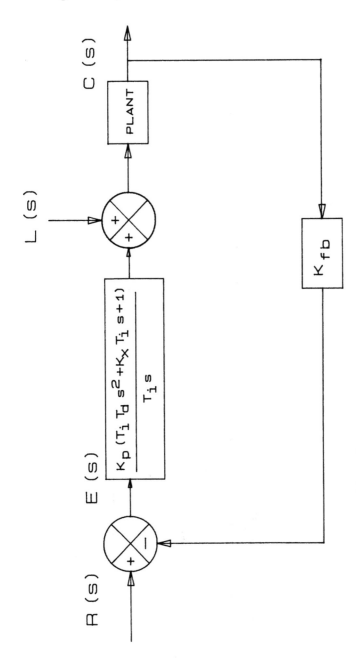

FIGURE 5.8 PID controlling plant.

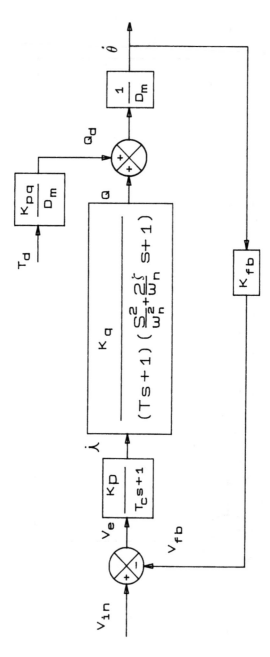

FIGURE 5.9 Speed-control with pseudo-integrator controller.

transducer. Figure 5.10 is the frequency response of the servovalve and motor (the motor dynamics are essentially negligible).
The open loop is

$$K_0 = \frac{K_p K_q K_{fb}}{D_m}$$

It is desirable to keep the open-loop gain as large as possible to
maintain adequate response and system stiffness when subjected
to a load. The block diagram of the system with respect to load
disturbances is shown in Figure 5.11. The closed-loop portion
reduces to

$$\frac{\dot{\phi}(s)}{T_1(s)} = \frac{1/D_m}{1 + K_0 G_p(s)}$$

where

K_0 $= K_p K_q K_{fb}/D_m$ = open-loop gain

$G_p(s)$ = plant dynamics

If K_0 were zero, the overall gain would be $1/D_m$. If K_0 were infinite, the overall gain would be zero. The closed loop is reduced in amplitude by K_{pq}/D_m, which produces the disturbance
flow from the load torque.

As stated earlier, it is desirable in a servovalve to keep the
slopes (K_{pq}) of the load-flow curve small. The slopes do increase at higher load pressures, especially at the envelope of
the curve where the maximum flow and pressure coexist. It is
highly desirable to keep away from this region in order to keep
the servosystem stiff (free from drooping due to loading conditions).

Therefore to keep loading effects from hindering servo performance, try to obtain high values of K_0, by compensation if
necessary, and to strive toward ideal low slopes for the load-
flow curves (small values of K_{pq}). As shown in Chapter 3, the
steady-state error of a system without an integrator is $1/(K_0 + 1)$.
Therefore a large open-loop gain will reduce the offset; the system will then approach the performance of the integrator, as
shown by these plots.

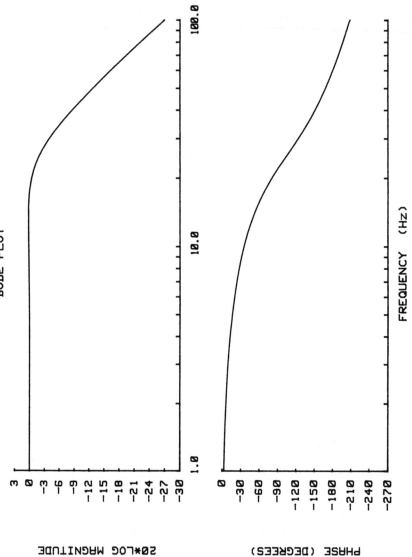

FIGURE 5.10 Open-loop frequency response of Figure 5.9 without lag controller.

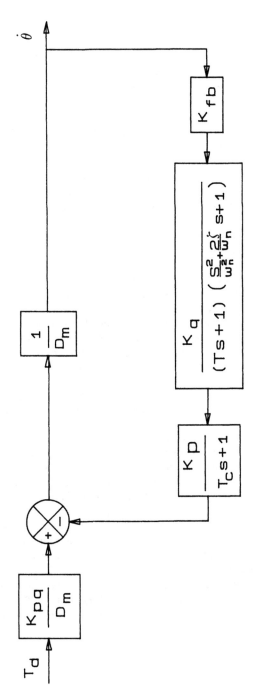

FIGURE 5.11 Flow disturbance to Figure 5.9.

Figure 5.12 is a plot of the open-loop response without the lag compensation. The open-loop gain is low. Figure 5.13 is a plot of the open loop with the compensated lag included. The open-loop gain has been raised considerably with adequate phase and gain margins. Figure 5.14 shows the closed-loop response for the compensated loop. Note that the peaked response without the compensator has been attenuated by the lag.

Since there is no integrator, the static (or low-frequency) gain is below 0 dB (that is, the output is less than commanded). With disturbances and loading effects, this offset will change even more. An integrator will eliminate the offset or droop of a proportional loop.

The lag compensator with high loop gain has brought the system close to an integrator-style loop. Figure 5.15 is the same closed-loop result with the first-order compensator, but with a closer look at the 0-dB magnitude range. There is an offset which may or may not be objectionable, depending upon the velocity requirements of the system.

If the system were approached with the PID compensator, it would be more expensive and subject to noise amplification of the feedback transducer (creating a need for a more expensive transducer). Because of the large value for the time constant, the first-order lag compensator acts like an integrator, but without the immediate phase loss of 90°. Figure 5.16 is the closed-loop response, with an integrator in place of the first-order lag (with a 0.8-s time constant).

It is easier to analyze and slow down a system to add stability than it is to add compensation to improve response and stability. It is important to fit the system to the requirements. In addition to compensation in the forward loop, compensation can be added in the feedback loop, with loop enhancements similar to the velocity inner-loop feedback.

The block diagram of Figure 5.17 is a system which uses a first-order lead in the feedback. Figure 5.18 is the same system reduced by removing the dynamic term from the feedback. Its enhancement comes in the form of a lead-lag. The lead is essentially within the loop; this increases the bandwidth of the plant by allowing a larger gain with the phase lead. The lag comes after the loop is closed and decreases the peaking due to low damping.

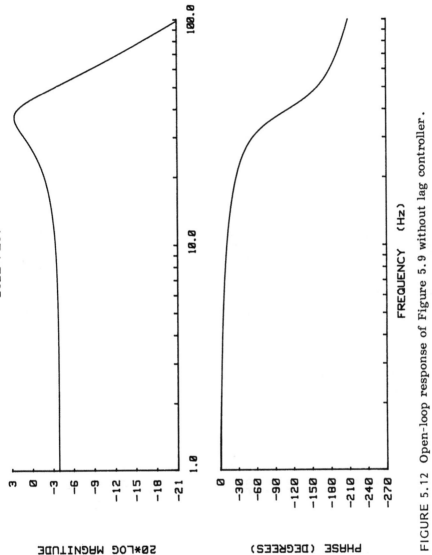

FIGURE 5.12 Open-loop response of Figure 5.9 without lag controller.

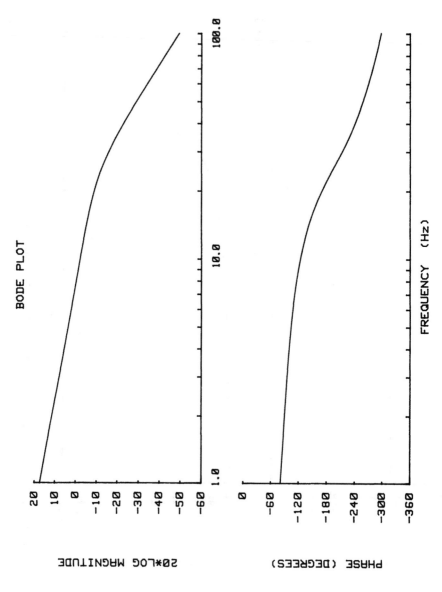

FIGURE 5.13 Open-loop of Figure 5.9 including lag controller.

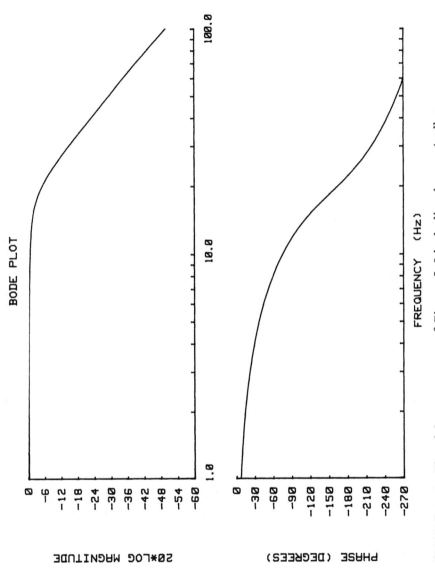

FIGURE 5.14 Closed-loop response of Figure 5.9 including lag controller.

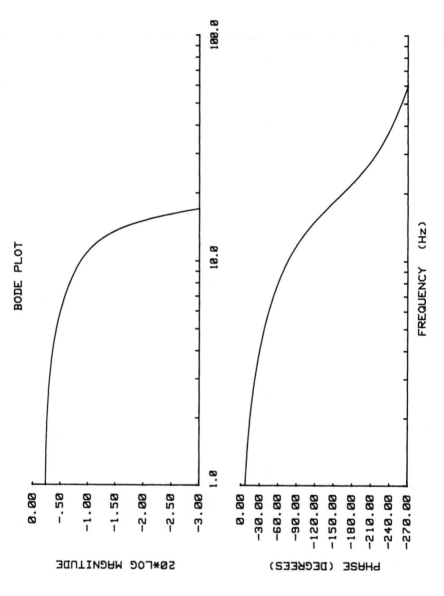

FIGURE 5.15 A closer look at the droop due to the pseudo-integrator.

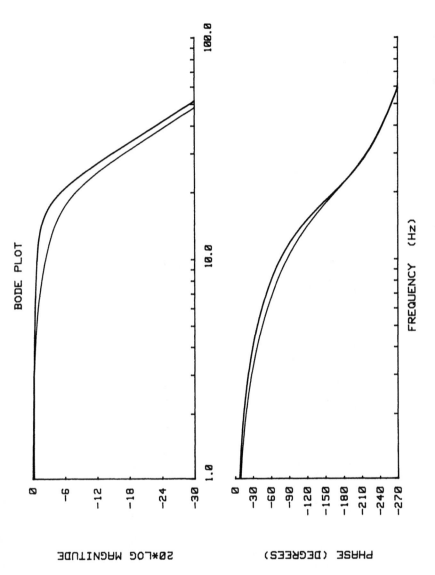

FIGURE 5.16 Integral control (lower curves on magnitude and phase) superimposed on the lag controller.

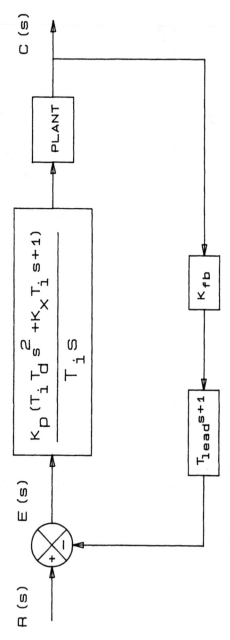

FIGURE 5.17 PID with lead in feedback.

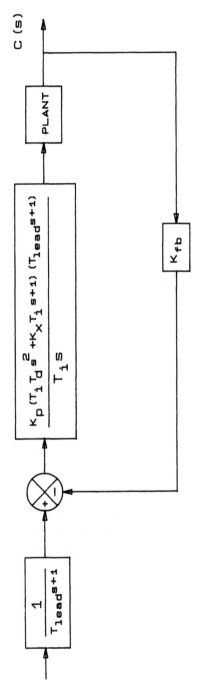

FIGURE 5.18 Equivalent block diagram of Figure 5.17.

215

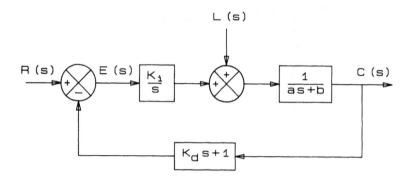

FIGURE 5.19 Integral controller with lead in feedback.

5.5 PSEUDO-DERIVATIVE FEEDBACK

Another control scheme, which has been coined "pseudo-derivative feedback" control by Richard Phelan [1], actually combines the feedback lead with the forward-loop integrator; this has some advantages for tuning the controller to the plant. Figure 5.19 is a system which has an integrator as the forward-loop element with a first-order lead in the feedback. The operational equation is

$$C(s) = \frac{K_i}{as^2 + (b + K_iK_d)s + K_i} R(s) + \frac{s}{as^2 + (b + K_iK_d)s + K_i} L(s)$$

Block-diagram reduction of the feedback results in the inner-loop feedback equivalent of Figure 5.20. If the integrator is placed before the inner loop, the diagram reduces to Figure 5.21. If K_i is put into the outer loop instead of the inner loop, the system configuration is not changed; only a multiply is relocated. This makes implementation of optimization of parameters easier. The resulting system is the pseudo-derivative feedback control scheme. It is shown in its usual form in Figure 5.22. The operational equation becomes

Inner loop

$$T.F._i = \frac{1/(as + b)}{1 + K_{d_1}/(as + b)} = \frac{1}{as + (b + K_{d_1})} \quad \begin{array}{l} \text{for the input R(s)} \\ \text{and L(s)} \end{array}$$

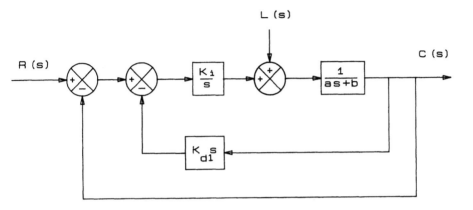

FIGURE 5.20 Parallel feedback path equivalent of Figure 5.19.

Outer loop

$$\frac{C(s)}{R(s)} = \frac{K_i/s(as + (B + K_{d_1}))}{1 + K_i/s(as + (B + K_{d_1}))} = \frac{K_i}{as^2 + (B + K_{d_1})s + K_i}$$

$$\frac{C(s)}{L(s)} = \frac{s}{as^2 + (b + K_{d_1})s + K_i}$$

$$C(s) = \frac{K_i}{as^2 + (b + K_{d_1})s + K_i} R(s) + \frac{s}{as^2 + (b + K_{d_1})s + K_i} L(s)$$

The pseudo-derivative feedback has several advantages. Unlike the PID algorithm, the pseudo-derivative scheme allows easier tuning because the terms K_i and K_{d_1} are independently adjustable to fit, respectively, the natural frequency and the damping ratio of the second-order lag. The block-diagram rearrangement has eliminated the need to physically locate the derivative portion of the lead in the feedback path; this makes the system less expensive, easier to implement, and less sensitive to noise. Because the block diagram was reduced to this form, the scheme is called "pseudo-derivative feedback."

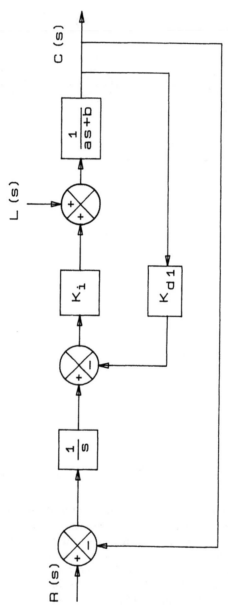

FIGURE 5.21 Movement of integrator to outer loop.

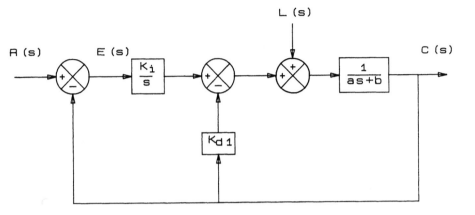

FIGURE 5.22   Pseudo-derivative controller driving first-order system.

If we add a second-order derivative to the inner-loop feedback, we get the block diagram of Figure 5.23; it has the operational equation

$$C(s) = \frac{K_i R(s) + sL(s)}{(a + K_{d2})s^2 + (b + K_{d1})s + K_i}$$

Multiplying numerator and denominator by $a/(a + K_{d2})$, we get

$$C(s) = \frac{\{a/(a + K_{d2})\}\{K_i R(s) + sL(s)\}}{as^2 + \{a/(a + K_{d2})\}(K_{d1} + b)s + \{a/(a + K_{d2})\}K_i}$$

which reduces to

$$C(s) = \frac{K_i' R(s) + \{a/(a + K_{d2})\}sL(s)}{as^2 + K_{d1}' s + K_i'}$$

This equation reduces to the normal form for the input R(s), but the input L(s) is reduced in amplitude by the ratio $a/(a + K_{d2})$.

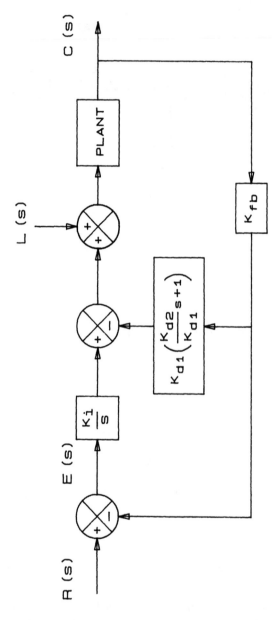

FIGURE 5.23 Second-order pseudo-derivative controller.

Therefore, the input transfer function is unchanged with K_{d2} present, but the load input portion of the operational equation reduces the effect of the load disturbance with larger values of K_{d2}. In reality, the effects of K_{d1} and K_{d2} are not quite so complete, because systems typically are larger than first and second order.

Figure 5.24 is a second-order pseudo-derivative loop with a third-order plant. The operational equation reduces to

$$C(s) = \frac{K_i R(s) + sL(s)}{as^4 + bs^3 + (c + K_{d2})s^2 + (d + K_{d1})s + K_i}$$

Obviously, the higher-order pseudo-derivatives would help shape the dynamic terms. However, physical implementation of a second- higher-order derivatives is difficult, mainly because of the amplifications of a potentially noisy signal. The scheme fits well for first- and second-order plants or systems which can break up into cascade sections of pseudo-derivative controllers.

This control scheme has a theoretical advantage in that the inner loop can be optimized first, with K_{d1} and K_{d2}. Then the outer loop can be adjusted for the proper value of K_i. In practice, this tuning approach can be used, but inner-loop optimization is not straightforward unless it is physically tested as only the inner loop. If the system were tested with both loops and small K_i, the response would be poor and mask the importance of the inner-loop sizing.

The PID and pseudo-derivative schemes allow a mechanism for tuning the compensation in a closed-loop system where the plant dynamics require more than just the proper gain for adequate response and stability. Although the integrator is essential for steady-state zero offset in the closed-loop output, it causes the response to suffer. If a parallel path for the input is placed at the summing junction of the pseudo-derivative feedback term, the configuration of Figure 5.25 results. Block diagram reduction back to the first summing junction has the effect of two parallel paths at the input, as shown in Figure 5.26. The result is a first-order lead at the input, given by

$$\frac{R_x(s)}{R(s)} = \frac{K_{ff}}{K_i} s + 1$$

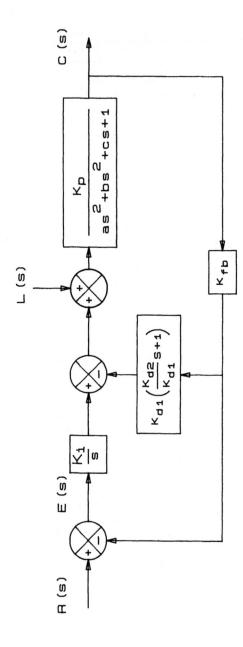

FIGURE 5.24 Second-order pseudo-derivative driving second-order plant.

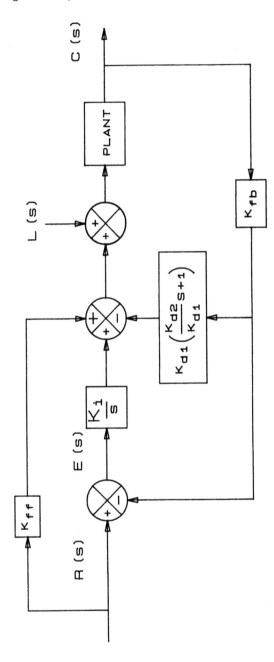

FIGURE 5.25 Pseudo-derivative controller with feedforward control.

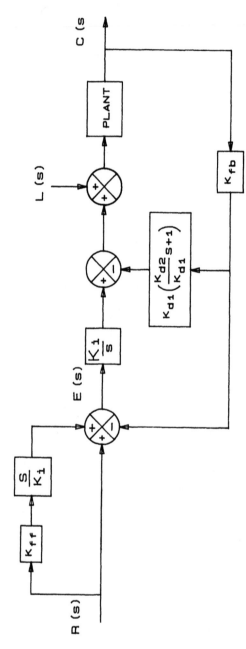

FIGURE 5.26 First-order lead equivalent of feedforward element of Figure 5.25.

The feedforward gain (K_{ff}) adjusts to fit the lead to the loop as part of the lead's time constant. Just as in the implementation of the pseudo-derivative feedback terms, the feedforward lead was accomplished without actually designing a derivative to obtain the lead. Note also that the feedforward helps the loop act quickly around the integrator in responding to abrupt input changes.

5.6 LEAD-LAG

The lag compensator presented in Section 5.4 utilized a first-order lag to perform the function of an integrator in obtaining little to no droop in output for a given command. It was also used to stabilize the plant dynamics in a closed-loop operation. If integration is not desired, or if the desired amount of the phase lag introduced by the compensator is less than the first-order lag, additional circuitry can be used. If it is desired to enhance the plant's response (but not as much as the PID, for stability reasons), combinations of lags and leads should be used.

Often the PID and pseudo-derivative-compensating schemes are insufficient if the system demands different output requirements. Often, in systems such as position or pressure control, the integrator is typically located at the plant's output as dictated by the feedback. Therefore, the PID and pseudo-derivative-feedback algorithms are not used because of the instability associated with this integrator.

A system may desire a rapid change in output if the input takes on a continuing large change in a relatively short time. The same system may wish to ignore smaller abrupt changes. When a plant and its loop are matched to the compensator to fit a wide range of frequencies and to attenuate some signals while amplifying others, the compensation typically needs lead-lag dynamics.

When a highpass filter with phase lead is needed to improve the transient response (higher bandwidth), lead compensation is used. This has a Bode plot similar to Figure 5.27. The added phase is placed where required by the plant or loop dynamics in order to increase the bandwidth.

The magnitude also increases with frequency to increase the transient response. If there is excessive noise in the system, this magnitude increase will amplify it. Since the low-frequency gain is attenuated, additional gain within the loop is necessary.

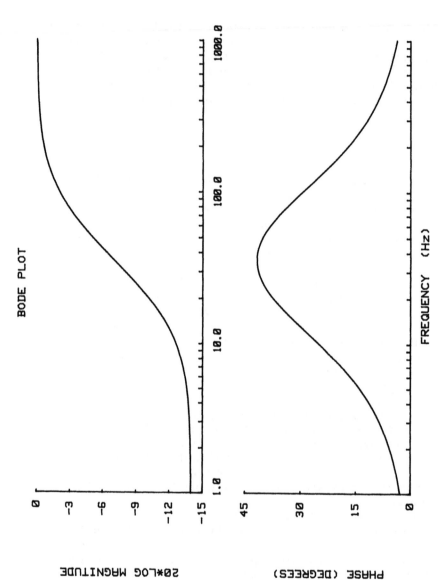

FIGURE 5.27 Lead-compensation dynamics.

The lag compensation of Figure 5.28 is used for low-frequency improvement of the steady-state error with high gains. The transient response of the resulting system is slowed down because it is a lowpass filter; this therefore creates a lower bandwidth system, as the gain crossover frequency is shifted to a lower frequency. If improvement in both the steady-state errors and transient response is desired, the lag-lead compensation of Figure 5.29 should be implemented.

The phase-lead portion adds lead and margin at the gain crossover frequency. The phase-lag portion places attenuation near and above the gain crossover frequency while allowing an increase in gain at low frequencies. These compensating techniques can be accomplished by digital and analog electrical circuits and by hydromechanical circuits. Figure 5.30 represents all of these techniques in a passive electrical circuit. Figure 5.31 is the impedance equivalent of Figure 5.30; the impedances Z_1 and Z_2 are defined as

$$Z_1 = \frac{1}{1/(1/C_1 s) + 1/R_1} = \frac{1}{C_1 s + 1/R_1} = \frac{R_1}{R_1 C_1 s + 1}$$

$$Z_2 = R_2 + \frac{1}{C_2 s} = \frac{R_2 C_2 s + 1}{C_2 s}$$

The circuit equations are

$$E_i(s) = I(s)\{Z_1 + Z_2\}, \qquad E_0(s) = I(s)Z_2$$

The transfer function of the output voltage (e_0) for the input voltage (e_i) is

$$\frac{E_0(s)}{E_i(s)} = \frac{I(s)Z_2}{I(s)(Z_1 + Z_2)} = \frac{Z_2}{Z_1 + Z_2} = \frac{(R_2 C_2 s + 1)/C_2 s}{R_1/(R_1 C_1 s + 1) + (R_2 C_2 s + 1)/C_2 s}$$

$$= \frac{(R_2 C_2 s + 1)/C_2 s}{[R_1 C_2 s + (R_2 C_2 s + 1)(R_1 C_1 s + 1)]/(R_1 C_1 s + 1)C_2 s}$$

$$= \frac{(R_2 C_2 s + 1)(R_1 C_1 s + 1)}{R_1 C_2 s + (R_2 C_2 s + 1)(R_1 C_1 s + 1)}$$

$$= \frac{(R_1 C_1 s + 1)(R_2 C_2 s + 1)}{(R_1 C_1)(R_2 C_2)s^2 + (R_1 C_1 + R_2 C_2 + R_1 C_2)s + 1}$$

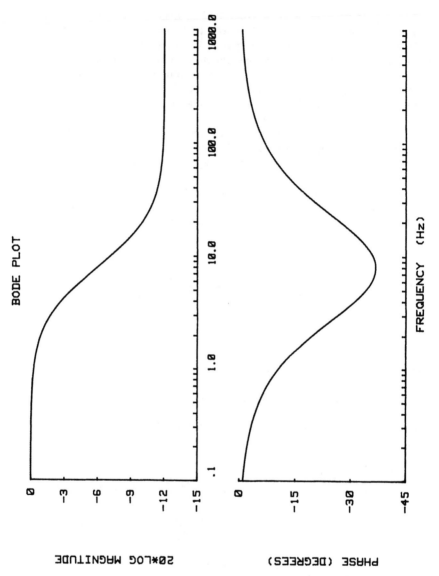

FIGURE 5. 28 Lag-compensation dynamics.

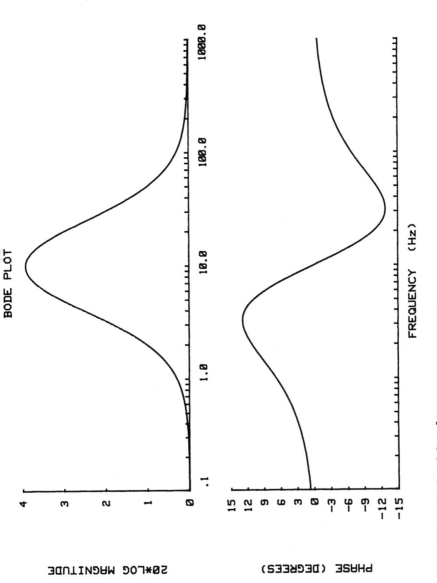

FIGURE 5.29 Lead-lag frequency response.

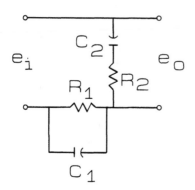

FIGURE 5.30 Electrical lead-lag passive circuit.

It is desirable to split the denominator into two first-order lags.
By inspection,

$$(\alpha T_1 s + 1)(\beta T_2 s + 1) = \alpha\beta T_1 T_2 s^2 + (\alpha T_1 + \beta T_2)s + 1$$

Defining $T_1 = R_1 C_1$ and $T_2 = R_2 C_2$, we obtain that the first- and
second-order constants equate to

$$\alpha\beta T_1 T_2 = T_1 T_2 \quad \text{and} \quad \alpha T_1 + \beta T_2 = T_1 + T_2 + R_1 C_2$$

$$\alpha = \frac{1}{\beta}, \quad \frac{T_1}{\beta} + \beta T_2 = T_1 + T_2 + R_1 C_2$$

With this substitution, the transfer function becomes

$$\frac{E_o(s)}{E_i(s)} = \frac{(T_1 s + 1)(T_2 s + 1)}{(T_1/\beta s + 1)(\beta T_2 s + 1)}$$

$$= \frac{(s + 1/T_1)(s + 1/T_2)}{(s + \beta/T_1)(s + 1/\beta T_2)} \quad \text{where } \beta > 1$$

Figure 5.29 reflects the magnitude and phase associated with
a particular choice of the time constants T_1, T_2, T_1/β and βT_2.
The low-frequency region portrays the lead-compensation circuit,
and the higher-frequency range represents the lag circuit.

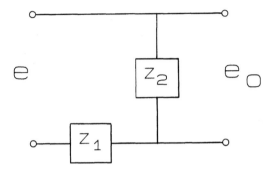

FIGURE 5.31 Impedance equivalent of Figure 5.30.

Besides being able to alter the inflection points, the circuit can be switched between lead-lag, lag, and lead by taking C_1 or C_2 out of the circuit. If a lag is desired, capacitor C_1 is removed; this yields the transfer function

$$\frac{E_o(s)}{E_i(s)} = \frac{Z_2}{Z_1 + Z_2} = \frac{(R_2 C_2 s + 1)/C_2 s}{R_1 + (R_2 C_2 s + 1)/C_2 s} = \frac{R_2 C_2 s + 1}{(R_1 + R_2) C_2 s + 1}$$

$$= \frac{R_2 C_2 s + 1}{[(R_1 + R_2)/R_2](R_2 C_2 s + 1)} = \frac{R_2 C_2 s + 1}{\beta R_2 C_2 s + 1}$$

$$= \frac{T_2 s + 1}{\beta T_2 s + 1} = \frac{1}{\beta} \frac{s + 1/T_2}{s + 1/BT_2}$$

where $\beta = (R_1 + R_2)/R_2 > 1$.

If a lead compensator is desired, the capacitor C_2 is bypassed. This produces the transfer function

$$\frac{E_o(s)}{E_i(s)} = \frac{Z_2}{Z_1 + Z_2} = \frac{R_2}{R_1/(R_1 C_1 s + 1) + R_2} = \frac{R_2(R_1 C_1 s + 1)}{R_1 + R_2 R_1 C_1 s + R_2}$$

$$= \frac{R_2(R_1 C_1 s + 1)}{R_1 R_2 C_1 s + (R_1 + R_2)} = \frac{R_2}{R_1 + R_2}$$

$$\times \frac{R_1 C_1 s + 1}{R_2(R_1 + R_2)(R_1 C_1 s + 1)} = \frac{\alpha(T_1 s + 1)}{\alpha T_1 s + 1} = \frac{s + 1/T_1}{s + 1/\alpha T_1}$$

where $\alpha = R_2/(R_1 + R_2) < 1$.

Obviously $\alpha = 1/\beta$. Note that the normalized forms of the lead-lag and the lag circuits each have a static gain of unity, whereas the lead circuit has a static gain of α, which is less than unity. The frequency where the maximum phase lead occurs can be calculated by taking the derivative of the phase differences, setting it to zero, and solving for ω :

$$\phi_t = \arctan(T_1\omega) - \arctan(\alpha T_1\omega)$$

$$\frac{d(\phi_t)}{dt} = \frac{T_1}{1 + T_1{}^2\omega^2} - \frac{T_1\alpha}{1 + T_1{}^2\alpha^2\omega^2} = 0$$

$$\left(\frac{d(\arctan U)}{dt} = \frac{1}{(1 + U^2)}\frac{du}{dt}\right)$$

Solving for ω, we find that the maximum phase is

$$\omega_{max} = \frac{1}{T_1\sqrt{\alpha}}$$

The frequency at which the maximum phase lead (ϕ_t) occurs can be substituted into the phase equation to obtain the relationship for the maximum phase as a function of α:

$$\phi_m = \sin^{-1}\left(\frac{1 - \alpha}{1 + \alpha}\right)$$

The lag and lead circuits, PID, pseudo-derivative feedback, velocity feedback, damping techniques and dynamic feedback circuitry all have the common function of compensation; this is accomplished by altering the location of poles and zeros of the root-locus plots to maintain stability while striving for optimal response. Routh's stability criterion is used to determine if the characteristic equation crosses over into the right-half s-plane, without solving for the roots of the equation. Compensation is utilized to keep the root-locus plots of a system in the left-hand s-plane for stability.

In general, the addition of poles, such as integrators and first- and second-order lags, will pull a system's root-locus plot toward the right and lower its relative stability. Zeros or derivative terms and lead elements, such as the second-order lead of the

PID, move the root locus to the left; this increases the stability
of the system.

The measuring element in the feedback path can be made,
sized, or combined with other circuitry to lower a system's sus-
ceptibility to external changes and loading effects, and to reduce the dynamic lag in the system. Compensation with a digital system involves a software algorithm rather than hardware.
The analysis and design of digital filters (compensation) also
varies from analog controls. Digital simulation, which was introduced in Chapter 3, will be expanded in a format analogous
to the Laplace evaluation.

5.7 FREQUENCY RESPONSE IN THE W-PLANE

It would be desirable to use frequency-response techniques on
digital systems because of the insight gained in sizing an open-
loop system for closed-loop control. Stability studies in the s-
domain and in the Z-domain differ by their stability boundaries.
The s-domain frequency-response data for obtaining phase and
gain margins employ the imaginary axis as the stability boundary. The stability profile in the Z-domain is the unit circle. By
transforming the unit circle into yet another plane, consistent
with the s-plane stability boundary, we can use a new frequency-
response technique.

The transform of the z-domain into the W-plane is defined by

$$z = \frac{1 + W}{1 - W}$$

This transformation allows the imaginary axis of the W-plane to
correspond to the unit circle of the Z-plane. Solving for W, we
obtain

$$W = \frac{z - 1}{z + 1}$$

This transformation is shown by evaluating the profile of the unit
circle with rectangular coordinates. The radius of the unit circle
is

$$|z| = e^{j\phi} = e^{j\omega t}$$

Therefore an evaluation of this magnitude as a function of ω yields (where z is replaced by $e^{j\omega t}$)

$$W = \frac{z-1}{z+1} = \frac{e^{j\omega t}-1}{e^{j\omega t}+1} = \frac{e^{-j\omega t/2}}{e^{-j\omega t/2}} \frac{e^{j\omega t}-1}{e^{j\omega t}+1} = \frac{e^{j\omega t/2}-e^{-j\omega t/2}}{e^{j\omega t/2}+e^{-j\omega t/2}}$$

Noting that

$$\sin\left(\frac{\omega t}{2}\right) = \frac{e^{j\omega t/2}-e^{-j\omega t/2}}{2j} \quad \text{and} \quad \cos\left(\frac{\omega t}{2}\right) = \frac{e^{j\omega t/2}+e^{-j\omega t/2}}{2j}$$

$$\tan\left(\frac{\omega t}{2}\right) = \frac{\sin(\omega t/2)}{\cos(\omega t/2)} = \frac{e^{j\omega t/2}-e^{-j\omega t/2}}{e^{j\omega t/2}+e^{-j\omega t/2}} \frac{1}{j} = \frac{W}{j}$$

We observe that W becomes a function of ω on the imaginary axis of the W-plane. This function is

$$W = j \tan\left(\frac{\omega t}{2}\right)$$

which reduces to

$$W = \tan\left(\frac{\omega t}{2}\right)$$

for $j\omega$ the imaginary part of ω.

If we exploit the identities

$$\sin(a) = 2\sin\left(\frac{a}{2}\right)\cos\left(\frac{a}{2}\right) \quad \text{and} \quad \cos^2\left(\frac{a}{2}\right) = \frac{1+\cos(a)}{2}$$

$$\sin\left(\frac{a}{2}\right) = \frac{\sin a}{2\cos(a/2)}$$

then

$$\tan\left(\frac{\omega t}{2}\right) = \frac{\sin(\omega t/2)}{\cos(\omega t/2)} = \frac{\sin(\omega t)}{2\cos^2(\omega t/2)} = \frac{\sin(\omega t)}{1+\cos(\omega t)}$$

and

$$W = j\,\frac{\sin(\omega t)}{1+\cos(\omega t)}$$

Therefore the circumference of the unit circle of the Z transform
maps into the imaginary axis of the W-plane. Because of the correspondence between the s- and W-planes, the Routh stability
criterion also can be extended to the W-plane.

The Z transform, together with the W transform, allows the
system performance to be investigated, compensated, and implemented. The Z transform adds a digital view to an analog plant
(which is to be a part of the digital system). Addition of compensation to the controller can be accomplished by the Z transform, but its implementation is not straightforward. Extension
of the system onto the W-plane allows for controller sizing, with
methods similar to those used in the analog system. Once the
controller is sized, the system or the controller is transformed
back into the Z-domain.

The performance analysis can be carried on completely in the
W-domain, especially the procedure to find the compensator settings. Once the controller is set, it is transformed back into the
Z-domain for implementation on the microprocessor. If the entire system is transformed back onto the Z-domain, the controller portion must be separately identifiable (independently from
the plant). The entire system can be analyzed in the Z-domain
if desired, but it is essential that the controller be transformed
because its sizing determines the microprocessor's main role.

Just as the differential equations involving integrators, lags,
and leads were analyzed by frequency response (Bode plots) to
determine system performance and stability, the difference equation (transformed into a function of W) also can be analyzed
through frequency-response techniques. Figure 5.32 represents
an analog plant combined with a digital controller. The hold circuit effectively becomes part of the plant. The digital compensator is assumed to be the resultant of the equivalent analog system lag circuit. This equivalence stems from the W-domain, which
yields the lag (analogous to that of the s-domain):

$$G_c(s) = \frac{Ts + 1}{\beta Ts + 1} = \frac{s + 1/T}{\beta s + 1/T}$$

$$D_c(w) = \frac{1 + W'/W\omega_0}{1 + W'/W_{\omega p}}$$

where

$G_c(s)$ = analog lag compensator

$D_c(s)$ = digital lag compensator in the W-plane

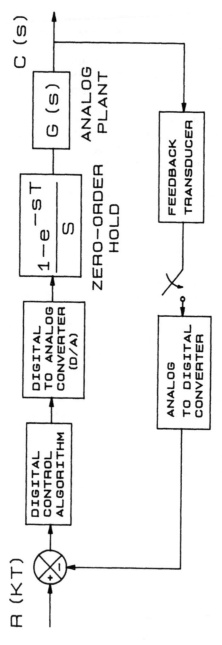

FIGURE 5.32 Digital controller with analog plant.

$W_{\omega 0}$ = "zero" location analogous to the zero $1/T$

$W_{\omega}p$ = "pole" location analogous to $1/\beta T$

Note that the zero and pole location frequencies W_{ω_0} and $W_{\omega}p$ are not equivalent to the corresponding frequencies (ω) of the analog system. They are related instead by $W' = j\tan(\omega t/2)$. By defining jW_ω as the imaginary part of W', we obtain

$$W_\omega = \tan\left(\frac{\omega t}{2}\right)$$

which relates the real (continuous) frequency (ω) to the W-plane frequency (W_ω).

The Z-transform equivalent of this W-plane digital compensator is obtained from the transformation $W = (z - 1)/(z + 1)$. This yields

$$D_c(z) = D_c(W)\Big|_W = \frac{z - 1}{z + 1} = \frac{1 + \dfrac{z - 1}{z + 1}\dfrac{1}{W_{\omega 0}}}{1 + \dfrac{z - 1}{z + 1}\dfrac{1}{W_{\omega}p}} = \frac{z + 1 + \dfrac{z-1}{W_{\omega 0}}}{z + 1 + \dfrac{z-1}{W_{\omega}p}}$$

The constants a, b, and c (b and c can be negative) reveal a similarity to the lag circuit defined both in the s- and W-planes (where Z is analogous to s and W). Once the system is evaluated and the compensator is sized for the proper gain settings in the W-plane, the Z transform provides the key to microprocessor implementation. Figure 5.33 represents an algorithm for the lag circuit.

Altering the orientation of $W_{\omega 0}$ and $W_{\omega}p$ (the zero and pole locations) can make the algorithm a lead circuit. First, the algorithm is set up to obtain the desired block diagram representing the lag (or lead) circuit. Then the equivalent assembly language (or higher-level language) is set up to fulfill the loop requirements. The left-hand branch, shown in Figure 5.33, has transfer function.

$$\frac{X(z)}{R(z)} = \frac{1}{1 + DZ^{-1}}$$

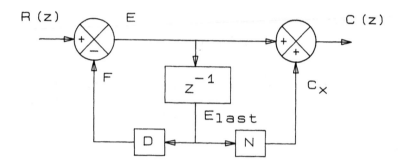

FIGURE 5.33   Lead-lag block diagram for microprocessor implementation.

The transfer function of the rightmost portion is

$$\frac{C(z)}{X(z)} = 1 + BZ^{-1}$$

Therefore the overall transfer function is

$$\frac{C(z)}{R(z)} = \frac{X(z)}{R(z)} \frac{C(z)}{X(z)} = \frac{1 + BZ^{-1}}{1 + DZ^{-1}}$$

The form (without Z notation) used to represent a first-order lag in Chapter 3 was

$$C_n = C_{n-1} + \frac{T_s}{T + T_s} (r - C_{n-1})$$

If this form is to be implemented on the microcontroller, then an algorithm can be established to implement it as a lead, lag, or lead-lag circuit. The total representation of the algorithm is shown in Figure 5.34 for an analog circuit. Note that by using the summations, two first-order leads and two first-order lags have been produced (by using only two first-order lags). The digital representation would be similarly implemented by replacing the lag elements $(T_1s + 1)$ and $(T_2s + 1)$ with C_1 and C_2 (outputs

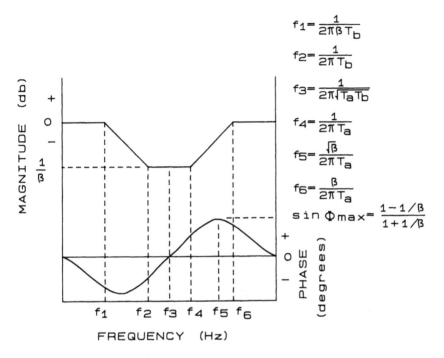

$$f_1 = \frac{1}{2\pi \beta T_b}$$

$$f_2 = \frac{1}{2\pi T_b}$$

$$f_3 = \frac{1}{2\pi \sqrt{T_a T_b}}$$

$$f_4 = \frac{1}{2\pi T_a}$$

$$f_5 = \frac{\sqrt{\beta}}{2\pi T_a}$$

$$f_6 = \frac{\beta}{2\pi T_a}$$

$$\sin \Phi_{max} = \frac{1 - 1/\beta}{1 + 1/\beta}$$

FIGURE 5.34 Lead-lag dynamic terminology.

of the digital first-order lag discussed in Chapter 3). The general form for the lead-lag circuit was (with the nomenclature of Figure 5.34)

$$\underbrace{\frac{T_a s + 1}{(T_a/\beta)s + 1}}_{\text{lead}} \quad \underbrace{\frac{T_b s + 1}{\beta T_b s + 1}}_{\text{lag}}$$

The lag is the right-hand section. The lead has a gain (less than unity) associated with it and is described as

$$\alpha \frac{T_a s + 1}{\alpha T_a s + 1} \quad \text{where } \alpha = \frac{1}{\beta}$$

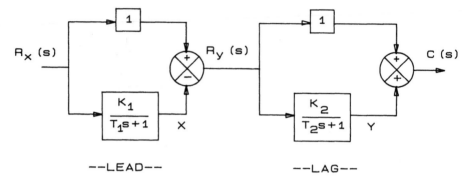

--LEAD--   --LAG--

FIGURE 5.35 Lead-lag algorithm.

The lag circuit (right portion of Figure 5.35) has

$$\frac{R_y(s)}{R_x(s)} = 1 + \frac{K_2}{T_2 s + 1} = \frac{\{1/(K_2 + 1)\}\{T s/(K_2 + 1)s + 1\}}{T_2 s + 1} \qquad \text{where } K_2 = \text{constant}$$

The lead circuit (left portion of Figure 5.35) has

$$\frac{C(s)}{R_y(s)} = 1 - \frac{K_1}{T_1 s + 1} = (1 - K_1)\frac{T_1/(1 - K_1)s + 1}{T_1 s + 1}$$

Cascaded together, they yield

$$\frac{C(s)}{R_x(s)} = \underset{1}{\frac{1 - K_1}{1 + K_2}} \ \underset{\text{lead}}{\frac{T_1/(1 - K_1)s + 1}{T_1 s + 1}} \ \underset{\text{lag}}{\frac{T_2/(1 + K_2)s + 1}{T_2 s + 1}}$$

In order to fit the general form, note that

$$\frac{1 - K_1}{1 + K_2} = 1 \qquad\qquad\qquad\qquad\qquad\qquad (1)$$

$$T_a = \frac{T_1}{1 - K_1} \quad \text{and} \quad \frac{T_a}{\beta} = T_1 \qquad\qquad\qquad (2)$$

result in

$$\beta = \frac{1}{1 - K_1}$$

$$T_b = \frac{T_2}{1 + K_2} \quad \text{and} \quad \beta T_b = T_2 \qquad (3)$$

result in

$$\beta = 1 + K_2 \quad \text{so} \quad \beta = \frac{K_2}{K_1} > 1, \quad \alpha = \frac{K_1}{K_2} < 1$$

Note that the lead circuit has the proper gain of α if it is to be used by itself. However, the lag implemented alone must have a static gain of unity. Therefore, the lag circuit must be multiplied by the inverse of its gain $(1 + K_2)$.

Making the algorithm operational on the microprocessor is accomplished by the flow of information as indicated in Figures 5.33 and 5.35. The assembly or higher-level language must perform the needed functions (a more detailed look at the microprocessor and its operating language is given in Appendix 2). Since $1/Z$ translates to the real-time unit-impulse function $\delta(t - T)$, its implementation on the microprocessor is the temporary storage of the variable or

Z^{-1} = temporary storage = time shift of one period

The Z-transform method of the lead or lag circuit in Figure 5.33 becomes, for example, in PL/M:

```
LEAD_LAG: PROCEDURE PUBLIC;
    E=R-D*Elast;
    C=Elast*N+E;
    Elast=E          /* update */
END LEAD-LAG;
```

In assembly language, the program flow in simplified form is

```
FIRST_ORDER:
    /* multiply the temporary storage (Elast) by the */
    /* value D (not from storage) and store into F    */
```

```
MUL     F,Elast,#D
/* subtract F from R and store into E */
SUB     E,R,F
/* multiply the temporary storage (Elast) by the */
/* value N (not from storage) and store into C   */
MUL     C,Elast,#N
/* add the value of E together with C and store in C */
ADD     C,E
/* update the storage register for the next call */
LD      Elast,E
END
```

The lead-lag compensation, without Z notation, would perform the same function of FIRST-ORDER (which was only a lead or a lag circuit depending upon A and B) if FIRST-ORDER's algorithm were increased to second order. The PL/M program of Figure 5.35 would be

```
MAIN: DO;
    DECLARE (-----,-----) BYTE
    DECLARE (-----,-----) WORD
    ----;
    ----;

    LAG:        PROCEDURE (OUT, LAST, IN, Kt) PUBLIC;
        OUT=LAST+Kt*(IN_LAST);
        LAST=OUT;
    END LAG;

    LEAD_LAG: PROCEDURE PUBLIC;
        CALL LAG(X,X_last,Rx,Kt);
        Ry=Rx-X*K1;
        CALL LAG(Y,Y_last,Ry,Kt);
        C=Ry+Y*K2;
    END LEAD_LAG;

    ------------;
    ------------;
    Kt=Ts/(Ts+T);
    CALL LEAD_LAG;
    ------------;
    ------------;
    END MAIN;
```

This method of compensation will be recapped. A system which contains a mixture of analog and digital controls must be properly matched, analyzed, and compensated (if required) to produce a responsive, stable system. First position(s) must be determined within the loop for the samplers. Then the pulse transfer function of the block diagram must be derived, including the hold circuits. The Z-domain can be transformed into the time domain, and parameters can be varied to check the performance.

The W-plane allows a graphical method of checking and compensating performance, by a procedure similar to that of the Bode plots in the s-plane. Therefore the plant is transformed from the Z-plane into the W-plane in order to study the plant and the open-loop dynamics for possible compensation. Then compensation circuits can be added. Once the compensator is sized for gain values, the compensator should be transformed into the Z-domain. See [2] and [3] for alternative digital filter designs.

Implementation of the compensator should be done by minimizing the number of total state times of the microprocessor. Compensation, when keyed to total system performance, results in improved system response. The open-loop gain, which will be shown to be indicative of the system's bandwidth, can be increased through compensation. Compensation, when properly employed, can aid in a system's interaction with the loading conditions.

5.8 CONCLUSION

With compensation networks available to reinforce the hardware's goal of optimal performance, the system design can be pursued. Even an ideal component of the system must be matched to other elements within the loop, including a feedback mechanism. Imperfections arise from loading effects, external disturbances, and noisy signals (especially those generated in the feedback path). The system becomes even more critical when the output requires stable, high response. The total system, whether all-hydraulic, electrohydraulic, or analog or digital, can be properly designed, compensated, and optimized prior to its physical construction.

BIBLIOGRAPHY

1. Phelan, Richard M. *Automatic Control Systems*, Cornell University Press, 1977.

2. Kuo, Benjamin C. *Digital Control Systems*, Holt, Rinehart, and Winston, 1980.

3. Ogata, Katsuhiko. *Discrete Time Control Systems*, Prentice-Hall, 1987.

6

Complete Systems under Control

6.1 INTRODUCTION

From simple to complex control schemes, the controller produces
an output (or set of output conditions) satisfying static profile
demands while maintaining stability and error signal optimization.
These systems vary from classical single-input, single-output
loops to coupled, multiple-input, multiple-output systems. Re-
gardless of the control scheme, the plant, or main item(s) to be
controlled, varies from system to system (and even within a
plant's operating range). Even the simpler plants can change
(through environmental and loading effects), depending on the
intent of the control scheme and operating range. Figure 6.1
is an example of a simple system with an open-loop gain K_O and
integrator $1/s$. The closed-loop equivalent is

$$T.F. = \frac{K_o/s}{1 + K_o/s} = \frac{1}{(1/K_o)s + 1} = \frac{1}{T_1 s + 1} \quad \text{where } T_1 = \frac{1}{K_o}$$

Figure 6.2 is a plot of this first-order lag (for a gain $K_O = 100$).
By definition, the corner, or cutoff, frequency f_c indicates the
bandwidth of the system (or component). The cutoff frequency
is the frequency where the amplitude is reduced by -3 dB. For
the simple first-order equation, the corner frequency occurs at
$f_c = 1/2\pi T$. Since this system constant is the inverse of the
open-loop gain, the cutoff frequency is

$$f_c = \frac{K_o}{2\pi}$$

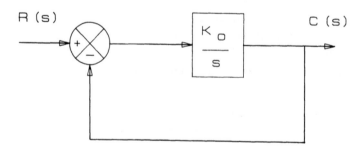

FIGURE 6.1 First-order control loop.

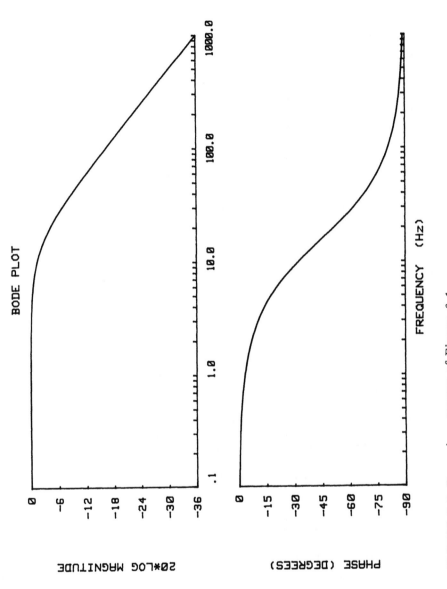

FIGURE 6.2 Dynamic response of Figure 6.1.

This implies that the open-loop gain can be increased indefinitely to produce a higher-response system. In reality, this is impossible because the assumption that the plant was behaving as an integrator could only be an approximation. Increasing the open-loop gain could move the system into a frequency range not studied or overlooked; this could easily reveal other dynamics which were negligible at lower frequencies. Noise in the feedback path and loading situations can also change the limits of the open-loop gain. If the increased open-loop gain brings about a "new" first-order lag $1/(T_1 s + 1)$, the closed-loop transfer function becomes

$$T.F. = \frac{K_o/s(T_2 s + 1)}{1 + K_o/s(T_2 s + 1)} = \frac{1}{(T_2/K_o)s^2 + (1/K_o)s + 1}$$

where

$$\omega_n = \sqrt{\frac{K_o}{T_2}}$$

$$\zeta = \frac{\omega_n}{2K_o} = \frac{1}{2\sqrt{K_o T_2}}$$

The natural frequency ω_n essentially represents the corner frequency or bandwidth of the system (the corner frequency also depends on the actual value of the damping ratio). Thus, by changing to include this first-order lag, the effective bandwidth also becomes a function of the first-order time constant. A relationship between the time constant T_2 of the second-order system's first-order lag and the time constant T_C associated with the first-order approximation can be made:

$$f_c = \frac{K_o}{2\pi} \quad \text{and} \quad K_o = \frac{1}{T_1}$$

from the first-order approximation, and

$$f_n = \frac{\sqrt{K_o/T_2}}{2\pi}$$

second-order definitions with regard to the closed loop (using the first-order lag with time constant T_2)

If the bandwidths of the first-order approximation and the
second-order system are equated,

$$f_c = f_n, \quad \frac{K_o}{2\pi} = \frac{1}{2\pi}\sqrt{\frac{K_o}{T_2}}$$

and if the open-loop gain is to be equal in both situations, then

$$K_o = \frac{1}{T_2} = \frac{1}{T_1}$$

Therefore, K_o remains indicative of the system's bandwidth when
the system is increased to a second-order system. Theoretically,
the second-order system is stable for all values of its open-loop
gain. However, the same limits exist as for the first-order as-
sumptions. Once a system becomes a third order, it can, depending on pole-zero placements, become unstable with certain
values of gain.

As the order of the system increases, so does the variance of
the bandwidth with the open-loop gain. Higher-order systems
imply either potential improvements or instability (with respect
to closed-loop operation) depending on the matching of the components (especially the controller). A first-order system asymp-
totically approaches a level when subjected to a step input. It
cannot have an overshoot as a second-order system would. The
third-order system can be sized to have even a sharper corner
in responding to this input signal.

The higher-order systems have more potential control over
the shape of the output, but deviations can be disastrous. In
other words, high-order systems have a high degree of potential when properly controlled. The region of controllability, however, fades with an increase in the order of the system. In Chapter 1, the human element of feedback actually controlled velocity
and position. The pseudo-derivative feedback algorithm of Chapter 5 controlled two derivative terms of the output. The human
reaction time placed limits on the effectiveness for its use in higher-response systems, whereas higher than second-order pseudo-
derivative terms are difficult to obtain without noise problems.

The controller and its implementation must be approached with
foresight into plant variance. Plant variance can occur between
systems which appear to be identical, and it certainly can occur
on component-varied plants. Obviously, a controller driving a
small pump and motor will respond and have different limitations

than a large pump-motor combination. The limitations on the open-loop gain become more apparent for systems of order larger than 2.

Within this established envelope are the limitations of electrical, flow, stroke, and other saturations as well as the limitations of loading demands and external disturbances. If the plant is controlled by its normal output (e.g., motor control, with speed as its output), then the controller, as well as the output, takes on a different form than if the plant output were integrated (by action of the controller) through the feedback path. A linear actuator would produce velocity as its output, but a position transducer will convert (integrate) its output to a position.

The controller may require an integrator in the velocity loop for steady-state accuracy. In the position-control loop, the controller is more likely to exclude the integrator for stability purposes (since the position transducer already has performed the integration). The pseudo-derivative approach could be used with position control in order to obtain control over the first- and second-order terms.

The pseudo-loop integrator adds one more order to the system; This infers tighter stability limits. Whether the system is controlling a plant with position, speed, force, or pressure control, the controller and the overall system response must be compensated and matched to the plant and its loading effects to produce responsive, stable systems.

6.2 PRESSURE-RELIEF-VALVE DYNAMICS

Pressure control takes on many forms. The pressure-control pilot stage is itself a closed-loop system, with output differential pressure proportional to input current. Similarly, the pressure-control boost stage, together with this pilot, is a closed-loop pressure-control device. Basic to almost all hydraulic systems is the pressure relief valve. The relief valve is a simple, closed-loop device which controls the pressure in a given circuit.

Responsive, stable systems result from responsive, stable components. A main, and usually the first, element of a closed-center system is the relief valve; it is also used to establish the upper pressure limit for open-circuit systems. In addition to producing a pressure level, the relief valve must ensure responsive, yet stable, operating conditions.

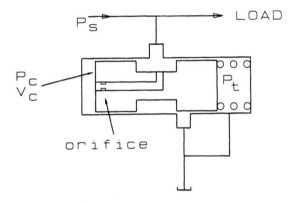

FIGURE 6.3 Relief-valve operation.

Whether a relief valve exists within or external to a loop, it must maintain an upper pressure limit. It can be analyzed from the physical laws involved in the flow and from those of a mass-spring arrangement. From the physical laws, the basic equations will be written for the relief valve of Figure 6.3. The equations will be placed into block-diagram form to show the resulting closed-loop system and to portray the interdependence of the component variables. Several time responses of typical configurations will be shown.

The equations of motion of the spool spring is

$$\Sigma F = ma = -Kx + (F_i - F_{fb})$$

where

F_i = input force = KX_c

F_{fb} = feedback force

X_c = input spool position and resulting spring preload

Rearranging and using the Laplace notation, we obtain

$$m\ddot{x} + Kx = F_i - P_c A = F_e(s)$$

$$X(s)[ms + K] = F_i(s) - AP_c(s) = F_e(s)$$

Since pressure is related to flow by

$$P = \int \frac{Q}{C_h}\, dt$$

$$P(s) = \frac{1}{C_h s}\, Q$$

$$Q = C_h sP(s) = \Sigma\, Q_i = \text{sum of flows}$$

Therefore the flow into the left end-control-chamber links the supply pressure to the control chamber pressure and spool stroke according to

$$\Sigma\, Q = Q_{fo}(s) + Q_x(s) = C_h sP_c(s) = \left(\frac{V_c}{\beta}\right) sP_c(s)$$

$$\text{where } Q_{fo} = Q_{orifice}$$

$$K_{pqf}[P_s(s) - P_c(s)] + AX(s) = \left(\frac{V_c}{\beta}\right) sP_c(s)$$

$$Q_x = \text{spool flow}$$

$$P_c(s) = \frac{AsX(s) + K_{pqf}P_s(s)}{(V_c/\beta)s + K_{pqf}} = \frac{(A/K_{pqf})sX(s) + P_s(s)}{(V_c/\beta K_{pqf})s + 1}$$

By substituting the control pressure P_c into the force equation, we establish a relationship between the stroke and the net force on the spool:

$$[ms^2 + K]X(s) = F_i - \frac{(A/K_{pqf})sX(s) + P_s(s)}{(V_c/\beta K_{pqf})s + 1} \quad (A)$$

$$X(s)\left[ms^2 + K + \frac{(A/K_{pqf})s}{(V_c/\beta K_{pqf})s + 1}\right] = F_i - \frac{AP_s(s)}{(V_c/\beta K_{pqf})s + 1}$$

$$X(s) \frac{[(mV_c/\beta K_{pqf} K)s^2 + (m/K)s + \{V_c/\beta K_{pqf} + A^2/K_{pqf} K\}s + 1]}{1/K\{V_c/\beta K_{pqf})s + 1\}}$$

$$= F_e(s)$$

where

$$F_e(s) = F_i(s) - \frac{AP_s(s)}{(V_c/\beta K_{pqf})s + 1}$$

$$\frac{X(s)}{F_e(s)} = \frac{1/K\{T_c s + 1\}}{as^3 + bs^2 + cs + 1}$$

where

$$a = \frac{mV_c}{\beta K \, K_{pqf}}, \quad b = \frac{m}{K}, \quad c = \frac{V_c}{K_{pqf}\beta} + \frac{A^2}{K_{pqf}K}$$

Figure 6.4 shows the relationship between the net force acting on the spool and the spool's resulting movement. The forces on the spool and the pressure in the control chamber are shown in block-diagram form in Figure 6.5. The flow pressure relationship is derived as follows:

$$\Sigma Q = C_h sP_s(s)$$

$$= Q_{in} + Q_{valve} - Q_{leak} - Q_{orifice}$$

$$= (Q_p - Q_1) + \{K_q X(s) - K_{pqm} P_s(s)\}$$

$$- K_1 P_s(s) - K_{pqf}\{P_s(s) - P_c(s)\}$$

$$= \frac{V_m}{\beta} sP_s(s)$$

$$
Fe(s) \quad \boxed{ \dfrac{\dfrac{1}{K}\left(\dfrac{V_c}{K_{pqf}B}s+1\right)}{\dfrac{m}{K_{pqf}KB}\dfrac{V_c}{}s^3 + \dfrac{m}{K}s^2 + \left(\dfrac{V_c}{K_{pqf}B} + \dfrac{A^2}{K_{pqf}K}\right)s + 1} } \quad X(s)
$$

FIGURE 6.4 Relief-valve force -- dynamic interaction.

Rearranging the terms as functions of control chamber, supply pressures, and the spool stroke, we obtain

$$
P_s(s)\left(\frac{V_m}{\beta}s + K_x\right) = (Q_p - Q_1) + K_q X(s) - K_{pqf}P_c(s)
$$

$$
= Q_p - Q_1 + K_q X(s) - K_{pqf}\left[\frac{P_s(s) + (A/K_{pqf})sX(s)}{\{V_c/\beta K_{pqf}\}s + 1}\right]
$$

and

$$
P_s(s)\left[\left(\frac{V_t}{\beta}\right)s + K_x\right] - \frac{K_{pqf}}{(V_c/\beta K_{pqf})s + 1}\,P_s(s)
$$

$$
= Q_p - Q_1 + \frac{K_q\{V_c(\beta K_{pqf})s + 1\} + AsX(s)}{(V_c/\beta K_{pqf})s + 1}
$$

and

$$
K_x P_s(s)\left[\frac{V_t}{\beta K_x}s + 1\right] - \frac{K_{pqf}}{(V_c/\beta K_{pqf})s + 1}\,P_s(s)
$$

$$
= Q_p - Q_1 + \frac{K_q\{(V_c/\beta K_{pqf})s + 1\} + AsX(s)}{(V_c/\beta K_{pqf})s + 1}
$$

This relationship is shown in block-diagram form in Figure 6.6. Figure 6.7, similar to [1], shows the complete relief-valve system.

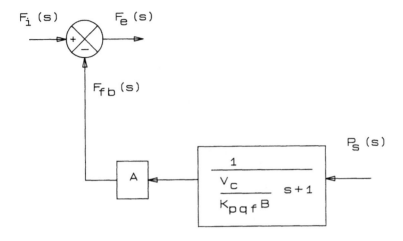

FIGURE 6.5 Control chamber feedback lag of relief-valve.

Figure 6.8 reflects the unity feedback form of the inner loop without load effects. Figure 6.9 is the unity feedback block diagram
of the load input without the setpoint input.

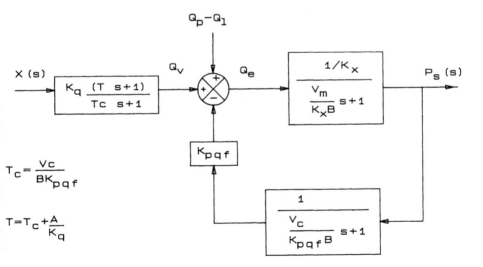

FIGURE 6.6 Inner-loop equivalent of relief valve.

256

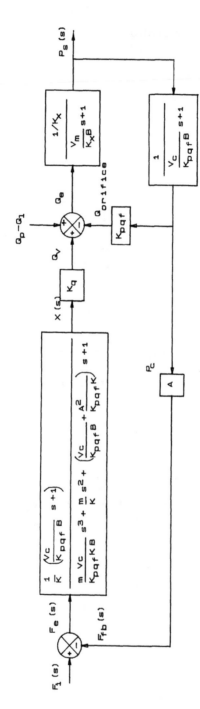

FIGURE 6.7 Relief valve control system.

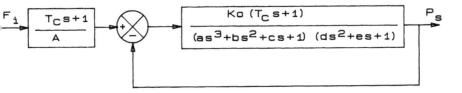

FIGURE 6.8 Relief valve results from setpoint input.

Superposition (in linear systems) allows us to evaluate both the setpoint input and the load input independently; these are summed to produce the output. Figures 6.8 and 6.9, which are the unity feedback equivalents resulting from setpoint and load inputs, respectively, will be evaluated for time-response comparison with variance in the control orifice (producing K_{pqf}) once the variables are defined.

In sizing a typical relief valve, assume it is desirable to flow 5 gpm at a maximum pressure rating of 2000 psi. In order to accommodate the 2000 psi pressure level concurrently with a maximum flow rate of 5 gpm, the relief valve must be sized to accommodate both the pressure and flow requirements. For a 0.25-in. diameter spool, the force requirements become

$$F_i = KX_i = \Delta PA = 2000(0.0491) = 98.2 \text{ lb}$$

In order to establish the proper combination of spring and stroke, the spool stroke is sized first to accommodate the travel requirements for flow; then the correction is added for the necessary preload of the spring. The resulting spring rate, together with the spool mass, must result in a natural frequency well above the dynamic response requirements of the relief valve and the controlled system. In order to port 5 gpm at the pressure setting of 2000 psi, utilizing a spool which ports 10% of its periphery, the resulting stroke is (from the flow equation)

$$X = \frac{3.85Q}{24.6\pi D \beta \sqrt{\Delta P}} = \frac{3.85(5)}{24.6\pi(0.25)(0.1)\sqrt{2000}} = 0.23 \text{ in.}$$

With the spool mass of 2.7×10^{-5} lb-s/in. and a desirable natural frequency of 700 Hz, the spring rate is

258

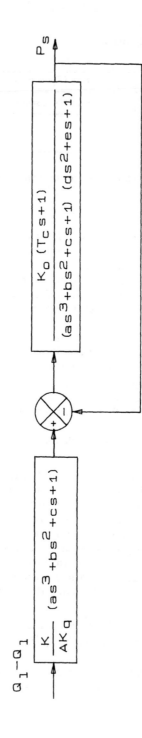

FIGURE 6.9 Load-flow input to relief valve.

$$F_n = \frac{\sqrt{K/m}}{2\pi}$$

$$K = (2\pi F_n)^2 m = (2\pi 700)^2 (2.7 \times 10^{-5}) = 522 \text{ lb/in.}$$

A spring rate of 500 lb/in. results in a spring preload of

$$X_i = \frac{F}{K} = \frac{98.2}{500} = 0.196 \text{ in.}$$

The spool travel for flow porting of 0.23 in., however, will cause inaccuracies because it takes up a big percentage of the preload stroke. Changing the porting periphery from 10 to 100% will decrease the porting stroke to a better match with the spring preload. With these values defined, the dynamic response can be implemented. Note that the input is set and effectively unchanged. The actual input is the change in flow the relief valve sees due to changes at the main load. The pump flow (Qp) to the load, if unrestricted, will equal the load flow (Ql).

When the load restricts the pump flow, the load flow becomes smaller than the flow provided by the pump; this creates a forced flow path at the relief valve. If the relief valve were an open port to tank, the combined flows of this passage with that of the load chamber would equal the input flow from the pump. Since the relief valve offers a flow restriction, this net input of pump flow minus load flow (Q_p - Q_l) represents the compressible flow (which builds up the supply pressure P_s to the static and dynamic characteristics described). Statically, the scale factor is set by the spool area, as indicated by the feedback path, or

$$\text{S.F.} = \frac{1}{A}$$

Within the loop, the porting stroke is minimized with respect to the preload stroke; this minimizes forced variations in the loop gains and therefore in the resulting output pressure. Since the system does not have an integrator, there is an offset from the desired output level. Once the loop is sized to minimize the output droop (striving for a large open-loop gain K_o), it must be analyzed for effectiveness against the input flow variations.

The following values will be used to study the relief-valve dynamics:

V_c = 0.0049 in.^{3} (control volume at orifice end of spool)

V_t = 4.7 in.^{3} (total volume of oil at input load to relief valve)

K_{pq} = 0.0029 cis/psi (slope of the load-flow curve of the main spool porting at a typical operating condition — actual from test data or estimated)

K_{pqf} = 0.0001 cis/psi (slope of the load-flow curve of the fixed orifice in the spool prior to the end control chamber)

K_q = 1600 cis/in. (slope of the output flow from the output flow orifice of the spool)

K_{leak} = 0.0001 cis/psi (leakage coefficient of the spool clearances)

β = 150,000 psi (bulk modulus of the oil)

A = 0.0491 in.^{2} (area of the end chamber of the spool)

m = 2.7 × 10^{-5} lb-s/in. (mass of the spool and one-third the mass of the spring)

K = 500 lb/in. (spring rate of the spring which biases the spool positioned into zero output flow to tank)

With these values defined, a comparison can be made between the time constants T and T_c. Note that T is actually equal to T_c plus A divided by K_q. Therefore T and T_c become

$$T_c = \frac{V_c}{\beta K_{pqf}} = \frac{0.0049}{(150,000 \times 0.0001)} = 0.00032$$

$$T = T_c + \frac{A}{K_q} = 0.147 + \frac{0.049}{1600} = 0.00032 + 0.000031 = 0.00035$$

Since T_c is approximately the same as T, they effectively cancel each other. By the superposition principle for linear systems, the loop can be analyzed independently for the output effects of both the main input and load input. The effect of the setpoint input is shown in Figure 6.10. Assuming the error signal of the main loop to be equal to zero, one can rearrange the loop as shown in Figure 6.9. With the values defined, the linearized load-flow step-response at setpoint input (load flow input) is shown in Figure 6.11.

If the fixed orifice is changed, which effectively produces a load-flow slope change, the step response changes with K_{pqf} as shown with various values (different orifice diameters). Note that the output is normalized to the setpoint input. This was

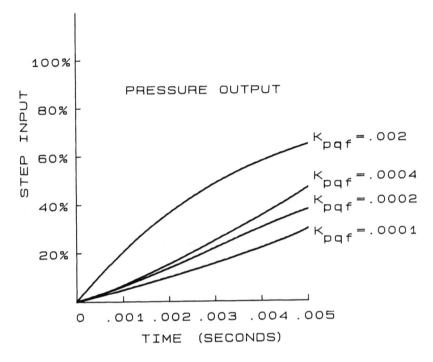

FIGURE 6.10 Setpoint input time response of relief valve.

done to represent the output of 2000 psi as 1 (or 100%). The scale factor of the setpoint input is $1/A$, and the load input is K/K_qA. Therefore, the scale factor of the load input equals the scale factor of the input multiplied by K/K_q or 500/1600, which is 31% of the input. The total result would be the sum of the setpoint input and the load-flow input.

Comparing these results with Figure 6.3 (the relief valve), we can gain insight into the valve's operation. First, the load influence will be discussed. If the fixed orifice is very small (resulting in a small K_{pqf}), the supply pressure (P_s) will build up very quickly, since it becomes difficult (timewise) to relieve itself to tank (because the spool is near its shutoff position). For larger-diameter orifices (larger K_{pqf}), the supply pressure has less restriction in accessing the control chamber at the end

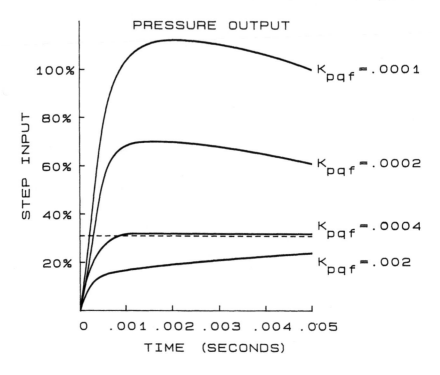

FIGURE 6.11 Load-flow input time response of relief valve.

of the spool. This results in a slower buildup of supply pressure.

The setpoint step responses indicate a quicker response with larger orifices (larger K_{pqf}). This is true because the spool can react more quickly in opening the spool. The combined effect of the setpoint and load determine the total output response of the supply pressure. There are obvious tradeoffs which must be matched by design to the type of circuit used and the accuracy required.

The relief valve must control difficult loading conditions. Sizing is critical. Stability must be maintained with minimal droop in set pressure. The state-space approach to control analysis (Section 6.5.1) can include the setpoint and load inputs simultaneously. Since the relief valve must handle a wide range of flows and load pressures, the nonlinear approach to state-space (Section 6.5.3) becomes an ideal simulation tool.

Frequency-response analysis becomes a good approach for sizing the controller to plant requirements. The following section uses several controller schemes sized to the plant through Bode (frequency-response) analysis.

6.3 VARIABLE-SPEED CONTROL

Speed-control devices can take on many forms. Valving can be utilized to control flow to a hydraulic motor, to obtain speed control of the shaft. An open loop could have a speed-control function. The feedback path becomes necessary to maintain the speed output with changes in load and environmental influences. The throttling losses of the valve(s) and the relief valve setting of the pump can make the system very inefficient. If response requirements don't dictate servovalve type of performance, then other, more efficient, methods of speed control should be implemented. Load-sensing mechanisms, as discussed in Chapter 2, could be used; this involves valving with a pump (with its controls) and motor.

If the pump is used within the loop and is directly attached to the motor, no valving between the pump and motor is required for motor movement; the pump performs only when required by the load and, therefore, by the loop. The speed-control loop shown in Figure 6.12 uses the pump within the loop. The pump control, as well as the controller and feedback mechanisms, could be totally hydraulic. Developments in producing electronics and electronic transducers suggest that the electrical controller is an effective means of closing hydraulic system loops.

In addition to cost incentives, the electrical controller, especially in digital form, opens up control schemes which were not previously possible. Each element of the plant represents some form of dynamic lag, which is compensated through feedback and the controller, to maximize and stabilize the requirements of the system. The components will be discussed and mated through block-diagram analysis. Finally, the system response and controller adjustments will be discussed.

6.3.1 Pump Stroker Control

In order for the pump to provide variable flow to the motor, either its input shaft speed must change or its volumetric flow rate must

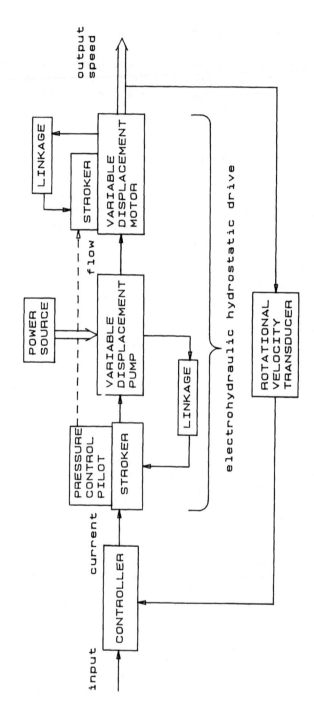

FIGURE 6.12 Speed-control loop.

264

change. Changing the input speed requires yet another control, with a resulting unresponsive system. As indicated in Chapter 2, the swash plate can be changed in its angular position to change the pump's output flow. Typically the swash plate is moved by using a piston arrangement (attached to the swash plate) which is driven by flow from a valving component.

The electrical stroke control shown in Figure 6.13 performs three functions: it provides the necessary flow to the swash-plate pistons to move the swash plate; it has an integral mechanical feedback with the flow-producing spool; and it beomes the controller for this inner loop. This inner loop provides a closed-loop position control of the swash plate for an electrical input command. The feedback allows the control to internally regulate swash-plate position without closing the control loop to other parts of the system.

The feedback link is similar in operation to that of the valve-ram application. For a temporarily fixed feedback position (indicating the present position of the swash plate), an input on the top of the linkage (from the pilot piston) will reflect through the feedback pivot to move the porting spool. This spool movement creates flow, in a four-way configuration, to the swash-plate pistons. As the swash plate moves, the linkage monitors this displacement through the drag link attached to the swash plate.

The feedback movement on the link is in the opposite direction to that created by the input from the pilot and piston. In the feedback mode, the pivot point on the link is at the input location (which is temporarily fixed at the command position). The feedback movement therefore forces the porting spool to move back toward its neutral position; this cuts off the supply of oil to the swash plate. With reduced, or smaller, perturbation flows at the swash-plate piston, the swash plate modulates through the closed-loop function of the feedback link and maintains a position proportional to the input.

The feedback linkage can be linearized as follows:

1. Hold X_{fb} constant (no feedback movement until the swash plate has moved by the porting spool flow)

$$x_p = L_a \theta$$

$$x_s = L_b \theta$$

266

Chapter 6

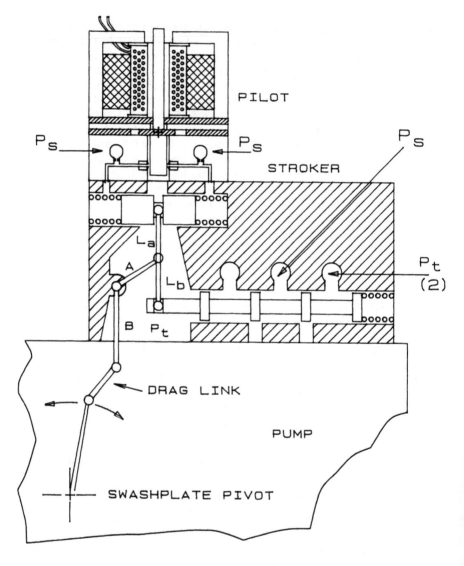

FIGURE 6.13 Electrohydraulic pump stroker control.

$$x_s = \frac{L_b}{L_a} x_p$$

where θ = angular movement about feedback pivot

2. Hold x_p constant (for a given input to the piston, the linkage is driven from the feedback to the porting spool)

$$x_s = (L_a + L_b)\theta$$

$$x_{fb} = L_a \theta$$

$$x_s = \frac{L_a + L_b}{L_a} x_{fb}$$

where θ = angular movement about piston input pivot

3. Combine input and feedback operation

$$x_s = \frac{L_b}{L_a} x_p - \frac{L_a + L_b}{L_a} x_{fb}$$

This result is shown in Figure 6.14 as the summing junction of the pump-stroker control.

The pilot stage is the pressure-control pilot stage discussed in Chapter 4. It provides a differential pressure output proportional to an input current. This differential pressure, working over the area of the piston ends, creates a force and moves the piston against the springs. When the spring's compression force balances the force from the pilot stage, the spool stops at a position proportional to the differential pressure (and therefore to the input current of the pilot stage).

The overall static function (scale factor) of the electrical stroke control is to provide an output pump position proportional to the input current. The block diagram of the pump stroke control is shown in Figure 6.14 (and reduced in Figure 6.15). As its modularity implies, the block diagram can be described as two cascade functions coupled through the capacitance of the piston variables (drive area and spring rate).

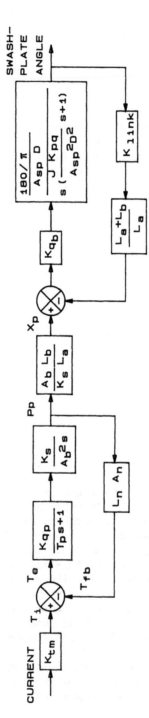

FIGURE 6.14 Block diagram of Figure 6.13.

268

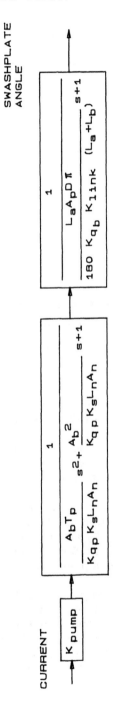

FIGURE 6.15 Reduced block diagram of Figure 6.14.

The pilot stage, as discussed in Chapter 4, has a static output of K_{tm}/L_nA_n psi/mA of input. Dynamically, the steepness of the load-flow curves (Figure 6.16) reflects that, for a given input, the resulting output will be maintained in the presence of disturbances (such as the feedback torques caused by the feedback motion of the linkage). The pilot will port more oil, with very little droop in output differential pressure, to counteract the disturbance and maintain the set differential pressure of the input. Thus the high slopes of the load-flow curve allow good impedance matching between the pilot and boost sections. The capacitance which the pilot sees is

$$C_h = \frac{A_b^2}{K_s} + \frac{V}{\beta}$$

where

A_b = drive area of the boost stage (piston)

K_s = spring rate of the boost stage (piston)

V = total volume driven by the pilot stage

β = bulk modulus of the oil

The larger the capacitance, the smaller the open-loop gain because the other loop gains remain constant (set by design). Therefore the larger the drive area or the smaller the spring rate, the less responsive the pilot and its load will become. The output differential pressure over the area of the piston, divided by the spring rate, creates a spool stroke proportional to the differential pressure.

This piston stroke works over the lever ratio L_b/L_a to produce the position of the porting spool. The porting spool (mathematically) has a flow gain K_q and creates flow proportional to its position. This boost-stage flow, divided by the area of the swash plate pistons, creates a piston velocity proportional to the porting spool position.

This velocity becomes an angular velocity, with respect to the swash-plate pivot. The feedback linkage "forces" the angular velocity into angular displacement. K_{link} is the gain of the linkage from the swash plate, through the drag link and feedback link to the floating link. The net result of the electrical stroke control is to produce adequate dynamic closed-loop control over

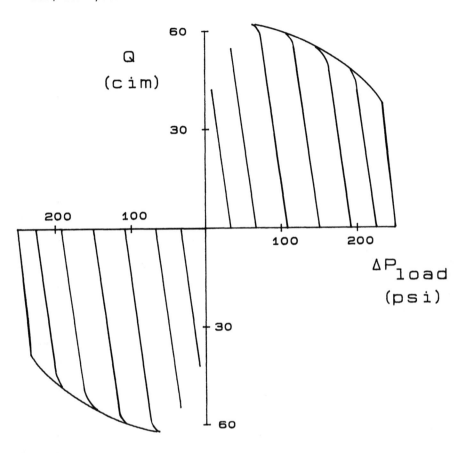

FIGURE 6.16 Pressure-control servovalve load-flow curve.

the pump's swash-plate displacement, with an adjustable dead-
band region for stable open-loop control. The adjustable dead
band allows for shifts in null output due to temperature, pressure, and tolerances involved in the piston, spool, and floating
link. A typical dynamic response of an electrical stroker control is shown in Figure 6.17.

6.3.2 Pump-Motor Dynamics

With the electrical stroker control providing pump displacement
proportional to input current, the pump is set to perform its

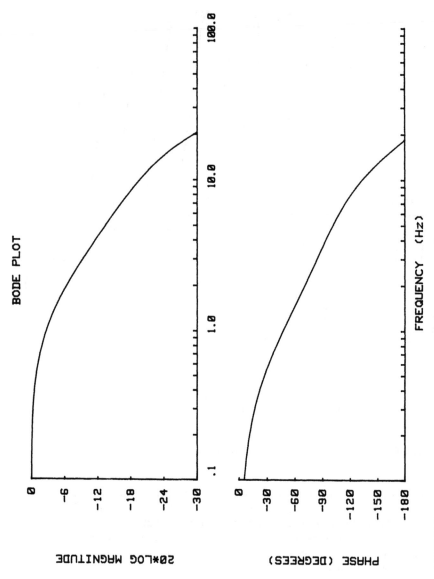

FIGURE 6.17 Frequency response of electrohydraulic pump stroker control.

function of providing proportional flow output. When the pump is matched to a motor, especially a variable motor, dynamic effects due to the pump and motor efficiencies, frictional damping at the load, and leakage play a major role in the resulting closed-loop response and stability. The variable motor introduces a control concern, because the varying control volume of the pump results in a change in the open-loop gain for both the main loop and an inner loop created by the pump-motor operation. The pump-motor operation will be described first in a block diagram, which allows a nonlinear study of the variables.

This style of nonlinear modeling can be handled by the analog computer or by the digital simulation programs available. The last portion of this chapter introduces the study of nonlinearities. A linearized form of the nonlinear pump-motor operation will provide a good means of analyzing the interaction.

Figure 6.18 is a nonlinear block diagram of the pump-motor stroker combination. The following derivation is based on the flow summation into the motor and torque summation at the output of the motor shaft. The compressible flow for a linear actuator was shown in Chapter 2 to provide the pressure to move the load. The motor is analogous to the actuator discussed, and causes a pressure buildup as a function of the compressible flow and leakage flow. The supply pressure is

$$P_s = \int \frac{1}{C_h} Q_c \, dt$$

which reduces, in Laplace notation, to

$$P_s(s) = \frac{1}{C_h s} Q_c(s)$$

where

$Q_c = Q_p - Q_l - Q_m$ = net compressible flow

Q_l = leakage flow in pump and motor = $K_l P_s(s)$

Q_m = motor flow resulting from motor output speed and motor displacement

P_s = resulting supply pressure in high-side line to motor

$C_h = V/\beta$ the hydraulic capacitance between the pump and motor

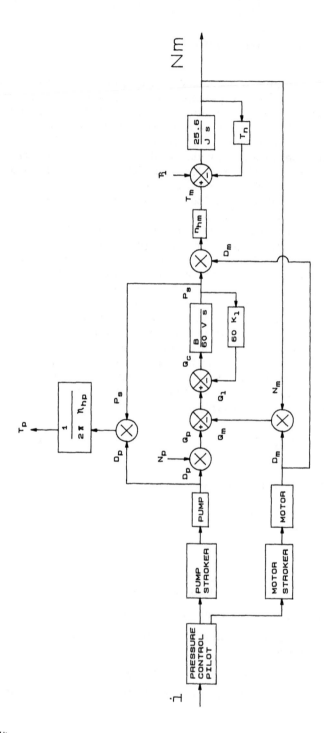

FIGURE 6.18 Nonlinear block diagram of Figure 6.12.

V = volume of oil in high-pressure line between pump and motor

β = the bulk modulus of oil

The pump flow minus the motor "feedback" flow(s) is essentially the compressible flow. The leakage presents a minor feedback loop. Block-diagram reduction of the supply pressure formulation (from the pump and motor flows) results from

$$P_s(s) = \frac{1}{C_h s} Q_c(s) = \frac{1}{C_h s} [(Q_p - Q_m) - Q_l] = \frac{Q_i(s) - K_1 P_s(s)}{(V/\beta)s}$$

$$P_s(s)\left[1 + \frac{K_1}{(V/\beta)s}\right] = \frac{1}{(V/\beta)s} [Q_i(s)]$$

$$\frac{P_s(s)}{Q_i(s)} = \frac{\beta/Vs}{1 + K_1\beta/Vs} = \frac{1/K_1}{(V/K_1\beta)s + 1} = \frac{1/K_1}{T_1 s + 1} \quad \text{where } C_h = \frac{V}{\beta}$$

Statically the output pressure is determined by the leakage coefficient. High leakage reflects low overall gain or low pressure output. The leakage coefficient is also built into the first-order lag. Since it is in the denominator, a large value of K_1 implies a small time constant. The small time constant is desirable to keep the lag effects minimal. Obviously there is a tradeoff for the desirable value of K_1. The loop will be reduced in the next section to show a clearer picture of the overall role of K_1.

The supply pressure (in psi) multiplied by the motor displacement (cis/psi) (reduced by the efficiency of the motor) is the torque input for the motor (T_i). This input torque must be sized to accommodate anticipated load torques (T_l) and losses at the motor. The sum of the torques on the motor shaft is

$$J_m\ddot{\theta} = J_m\dot{V}_m = \dot{N}_m J_m = T_i - F_n N_m - T_l$$

where

J_m = rotational inertial mass of the motor

N_m = motor speed

F_n = frictional damping coefficient proportional to velocity

T_l = load torque on the motor

In Laplace notation, this reduces to

$$J_m sN(s) + F_n N_m(s) = T_i - T_l$$

The feedback transducer measures velocity; therefore the torque summation yields an inner-loop first-order element. This can be seen by reducing the last equation to

$$N(s)(J_m s + F_n) = (T_i - T_l)(s) = T_e(s)$$

$$\frac{N(s)}{T_e(s)} = \frac{1}{J_m s + F_n} = \frac{1/F_n}{(J_m/F_n)s + 1} = \frac{1/F_n}{T_f s + 1}$$

The motor speed output, after being multiplied by the motor displacement, is the motor flow (as shown in Figure 6.18). The displacement motor setting is set by the stroker control on the motor. This "motor flow" is the "resistance" which builds up supply pressure from the available flow of the pump. If the load were such that the motor couldn't rotate, the pump would provide the flow against (the capacitance equivalent of) the entrapped oil to build up supply pressure to the pressure relief setting.

For effective unloaded motor operation and low operating speeds, the supply pressure buildup will be small because the effective motor "feedback flow" is close to the value of the pump flow (because the motor is, in effect, unrestricted). When the motor speed is high yet unrestricted (no output loading), the motor flow and pump flow also become essentially equivalent, with little resulting pressure buildup.

Figure 6.18 contains the variables of the pump and motor and their stroker controls. Note that the multiplication junctions means that the simulation is nonlinear. With loading conditions on the output motor shaft, any speed condition, especially high speed, will cause the "motor feedback" flow to be less than the pump input flow; this will build up the supply pressure. The net flow resulting from the pump and motor instantaneous conditions works against the capacitance of the high-pressure side (between the pump and motor). Statically, the resulting pressure from the pump-motor compressible flow is determined by the leakage coefficient (K_l) of the pump and motor.

If the motor displacement is decreased, the resulting torque capabilities is diminished and the motor feedback flow is decreased;

this results in a pressure buildup. The smaller motor displacement results in higher output speeds. The torque, produced
by the supply pressure buildup, and the motor displacement are
sufficient to overcome the inertia of the motor. The inner loop
at the torque summation has a scale factor of $1/F_n$, where F_n is
the frictional damping coefficient. Statically, an increase in
frictional damping results in lower output speed, just as would
be expected for a load disturbance.

The frictional damping is also present in the time constant (T_f)
of the first-order lag. An increase in this friction results in the
desirable decrease in T_f. Just as the leakage coefficient has a
role in transforming flow into pressure, the frictional damping
is also present in the time constant T_f of the first-order lag.

The torque loading, due to the pump displacement (D_p) and
the resulting high pressure (P_s) on the pump input source, is

$$T_p = \frac{1.37 D_p P_s}{\tilde{n}}$$

where T_p is the resultant torque requirements on the input energy source to the pump and $\tilde{n}$ is the pump efficiency.

The motor stroker control can be identical to the pump stroker
control, except that it usually operates in only one direction. The
input to the stroker control could be a pressure-control pilot
stage, integral in its design, or it could use the pilot of the
pump stroker control. When the same pilot is shared between
the pump and motor, the pump is stroked first.

After the pump has reached its maximum swash-plate position, the differential pressure of the pilot stage is matched to the
proper working direction by a shuttle valve; this also provides
the differential pressure to the motor stroker. The motor will
start at a maximum displacement and decrease to a minimum setting. This is a good approach because the speed should be staged
to get the proper proportionality between output speed and input
current. If higher output supply pressures are desirable for effective low-load requirements (for stability or other reasons), independent pump and motor strokers may be beneficial.

6.3.3 Linearized Plant

Linearization of this block diagram gives more insight into the use
of the valve parameters for controllability. The motor feedback

flow equivalent can be analyzed by assuming a fixed motor. By running a simulation (including a frequency-response plot) at this displacement and comparing it with a simulation at another value of motor displacement, an alternative to physically testing a pump-motor combination is achieved. This would replace a frequency-response test, with an offset in the input to place the baseline of the test in the motor range.

The amplitude of the frequency-response input would be small to keep the operating point at the "fixed" motor displacement. The linearized equivalent block diagram, discussed in [2], is shown in Figure 6.19 with the swash-plate position of the pump as the input and motor speed as the output. The motor displacement sets the scale factor of the motor loop; the smaller the displacement, the higher the resulting speed. The motor displacement also affects the loop's internal characteristics.

The open-loop gain of the inner-loop motor contains the gain (D_m) of the motor displacement as a squared function. A small increase in D_m causes a much larger increase in the open-loop gain; similarly, a decrease in D_m causes a lower open-loop gain. Lowering D_m decreases the response of the inner loop; this makes it a more dominant pole (lag).

The reduced frequency, due to the response of the motor inner loop coupled with its effective increase in gain in the main loop, tends to bring the system toward instability (if it was set up to maximize the pump stroke range). In other words, increasing speeds by means of the variable displacement results in a lower stability margin under closed-loop operation. Figure 6.19 shows the reduced form of the two first-order lags within the motor loop. The first lag has a time constant

$$T_1 = \frac{V_s}{\beta K_1}$$

where

V_s = volume of oil on high side of pump and motor

β = bulk modulus of the oil

K_1 = the leakage coefficient of the pump motor combination (available from operating curves of particular pump and motor)

If K_1 increases, the time constant decreases; this makes the lag less of a hindrance in the total dynamic loop. However, as shown,

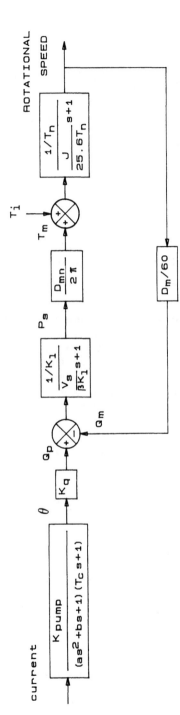

FIGURE 6.19 Linearized block diagram of Figure 6.18.

the open-loop gain is indirectly proportional to the leakage coefficient; this gives a lower open-loop gain, or it makes the inner loop of the motor more sluggish. The volume V_s should be kept to a minimum to reduce the lag effects associated with the resulting time constant T_1.

The second lag in the motor loop has a time constant

$$T_f = \frac{J_m}{25.6f_n}$$

where

T_f = time constant of the motor inertial damping

J_m = rotational inertia of the motor and output shaft

f_n = frictional (velocity dependent) damping

Obviously, a high inertia will slow the system response to input commands. The frictional damping tends to decrease the lag effect, especially when the output shaft is heavily loaded. The damping term F_n also appears in the open-loop gain of the motor loop; a large value of F_n actually makes the system less responsive. By reducing this motor loop's block-diagram form, the effect of each term will become more obvious. The reduced form is shown in Figure 6.20; this results in the transfer function

$$T.F. = \frac{K_o/(T_1s + 1)(T_fs + 1)}{1 + K_o/(T_1s + 1)(T_fs + 1)} = \frac{K_o}{(T_1s + 1)(T_fs + 1) + K_o}$$

$$= \frac{K_o/(K_o + 1)}{[T_1T_f/(1 + K_o)]s^2 + [(T_1 + T_f)/(1 + K_o)]s + 1}$$

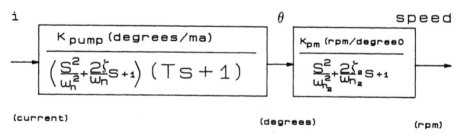

FIGURE 6.20 Reduced block diagram of speed-control loop.

Equating the standard second-order terms for natural frequency and damping ratio, we obtain

$$F_n = \frac{\omega_n}{2\pi} = \frac{1}{2\pi}\sqrt{\frac{1 + K_o}{T_1 T_f}} \quad \text{and} \quad \frac{2\zeta}{\omega_n} = \frac{T_1 + T_f}{1 + K_o}$$

$$\zeta = \frac{T_1 + T_f}{2\sqrt{(1 + K_o)(T_1 T_f)}}$$

If the open-loop gain (K_o) is low, the system will become sluggish. If $K_o \gg 1$, then the natural frequency (F_n) and the damping ratio (ζ) become

$$F_n = \frac{D_m}{2\pi}\sqrt{\frac{25.6\tilde{n}\beta}{2\pi 60 V_s J}}$$

$$\zeta = \left(\frac{V_s}{\beta K_1} + \frac{JK_1}{25.6f_n}\right)\frac{1}{D_m}\sqrt{\frac{2\pi 60\beta}{\tilde{n}V_s J}}$$

If $T_1 \gg T_f$, then

$$\zeta = f_n\sqrt{\frac{2\pi V_s 60}{\beta \tilde{n} J}}$$

If $T_f \gg T_1$, then

$$\zeta = \frac{K_1}{25.6 D_m}\frac{2\pi 60\beta J}{\tilde{n}V_s}$$

This gives a clearer look at the effective bandwidth of the motor loop (because the natural frequency of a second-order lag is indicative of the bandwidth): the larger the motor displacement, the larger the bandwidth. Therefore the motor dynamics will be quicker when at its maximum setting. When the motor is stroked toward its minimum setting, the system will be less responsive. Either high-side volume or the inertial mass, if increased, will slow the motor response or reduce the bandwidth.

High efficiencies and good values of bulk modulus (no air or
minimal hose effects) will keep the bandwidth high. Because of
the square root, the motor displacement becomes the dominant
variable in affecting response. Note that neither K_l nor F_n determines the natural frequency. The damping ratio isn't as clear,
unless one of the time constants is dominant.

If the time constant with the leakage (K_l) is dominant, then
the damping ratio is a function of F_n. Likewise, if the damping
time constant is dominant, then the damping ratio is a function
of the leakage coefficient. Note also that, depending on which
time constant is dominant, F_n determines the inverse relations
on the remaining variables of $\tilde{n}$, V_s, β, and J_m. The system
variables K_l and F_n do change in the operating regions of many
systems; this can put the system into marginal stability.

The controller used in a closed-loop system becomes the key
element for good speed control. Figure 6.21 is a frequency-response plot of a typically sized pump-and-motor combination (in
its pump stroking range) using the stroker of Figure 6.13. Figure 6.22 is a frequency plot of the same system in the motor range.
Both the pump and motor range are shown superimposed in Figure 6.23.

The controller can add phase lead for the plant, or it can provide a secondary means of controlling the damping. The terms
which affect the damping ratio but not the natural frequency are
K_l and F_n. It is difficult to control the velocity coefficient F_n because it is a function of the torque, speed, and loading effects.
Under severe changes in loading conditions, this velocity coefficient (F_n) may be the cause of low damping. It would be difficult
to control this term, because implementation would appear like
braking; this wastes energy and is difficult to resolve into controllable methods. Instead K_l must be raised to increase the damping ratio.

By measuring overspeed in the situation of low damping (due
to loading conditions) or by detecting the motor operating range,
we can determine how to increase the leakage coefficient in order
to increase the effective damping ratio. Increased leakage (by
valving) can also help the overall response when the system approaches marginal stability (pump-motor inner-loop gain change
as well as dynamic lag change). The motor stroking range input
signal (differential pressure) can be used to request more leakage.

An electrohydraulic two-way servovalve can be situated to
"bleed" the high-side pressure to the low side of the pump motor.
The same pressure-control pilot which drives the pump and motor
strokers can be used, if the source of low damping is caused by

the motor stroking range. In this case, the increase of input to the motor speed (obtained by decreasing its swash-plate angle) will be the proportional input to the two-way valve. Depending on the physical pump and motor sizes, the two-way bypass valve (which is not controlwise tied to the stroker controls) should have minimal leakage.

Direct-acting single-stage spool valves or jet-pipe pilots tend to have very low leakage when the bypass valve function is de-activated. The advantages of null adjustability, modularity, and minimal machining tolerances of the two-spool, four-way valve can be extended to the two-way valve bypass function.

The differential pressure-control pilot, remotely operated from the charge pump, fulfills the function of minimal standby leakage. The spool and body do not have any axial precision machining — just a single cleanup on the spool and body is needed to provide good meterability. The precision lap is done by the null adjustment. The valve would be powered up to its neutral, or closed output, condition to provide a fail-safe feature. Note that the pilot leakage is in another hydraulic circuit and does not affect the boost-stage leakage function.

6.3.4 Controller Operation

The controller, whether providing phase lead for the plant or sending the appropriate signal to the bypass valve, must be ca-pable of maximizing the scale factor (of speed output) for com-mand input under load variances, internal loop disturbances, and feedback noise effects. The loop will first be investigated within the realm of adding appropriate phase and gain margins to maxi-mize response with stability. This will be done with both the PID (proportional plus derivative plus integral) controller and a pseudo-derivative controller.

6.3.5 PID Speed Loop

Figure 6.24 represents the PID controller matched to the plant. The plots of the plant (pump, motor, and strokers) of Figures 6.17, 6.21, and 6.22 must be modified by an integrator to ensure zero offset to a step input. For this example, the feedback scaling of 60 rpm (60 teeth/s) per 5V of input reflects a gain necessary for the open-loop investigation.

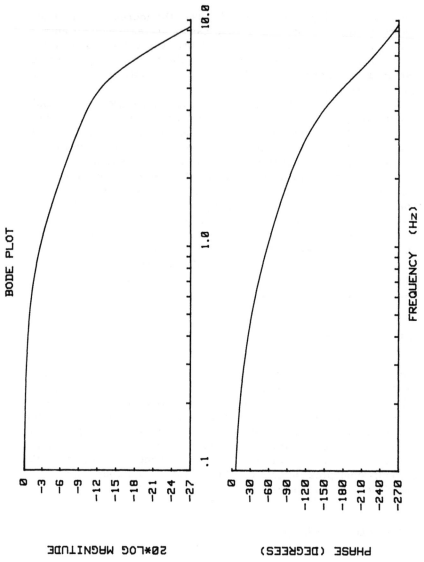

FIGURE 6.21 Hydrostatic transmission (coupled pump-motor) frequency response.

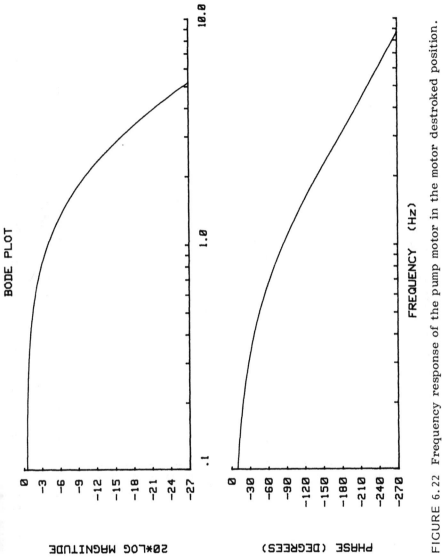

FIGURE 6.22 Frequency response of the pump motor in the motor destroked position.

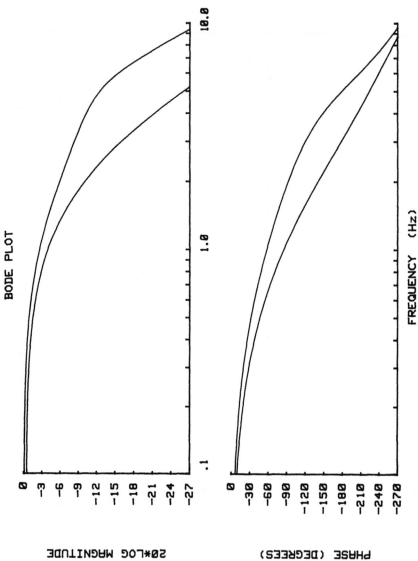

BODE PLOT

20*LOG MAGNITUDE

PHASE (DEGREES)

FREQUENCY (Hz)

FIGURE 6.23 Frequency response of pump range (faster response) superimposed with the

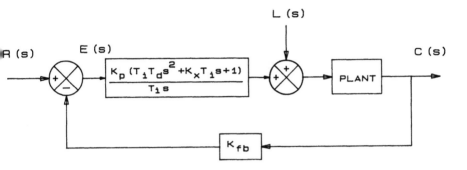

FIGURE 6.24 PID loop.

To determine the gain margin and phase margin of the system, we plot the static and dynamic plant dynamics, together with the static and dynamic parameters of the integrator and feedback elements, in the frequency domain. Figure 6.25 represents the resulting minimum plant controller requirements ($K_p = T_i = K_x = 1$ and $T_d = 0$). It is desirable to increase the open-loop gain as much as possible for the reasons given in Chapter 3 (basically, to minimize loading effects, increase response, and optimize error criterion).

Figure 6.26 indicates that the open-loop gain K_0 can obtain a value of 30 while maintaining a phase margin of 45° and a gain margin of 10 dB. Figure 6.27 is a root-locus plot of the system, with gain K_0 as the variable parameter (and $D_m = 10$ in.^{3}/revolution). Figure 6.28 is a root-locus plot with the motor displacement changed to $D_m = 5$. The maximum open-loop gain for stability is decreased by a factor of 3 when the motor displacement is decreased by a factor of 20. The integrator entered a destabilizing effect, which must be countered by compensation in order to effectively increase the response.

The PID is typically difficult to adjust. Also the PID algorithms usually use $K_x = 1$; however, some benefits result from using K_x as a variable. If T_i is normalized or set to a value of unity, the controller gain is a function of the desired proportional gain K_p. Also, lead-controller dynamic benefits result. If the derivative gain (T_d) is zero, the lead element is first order with time constant of $K_x T_i$. With T_i set to unity, the time constant is a function of K_x.

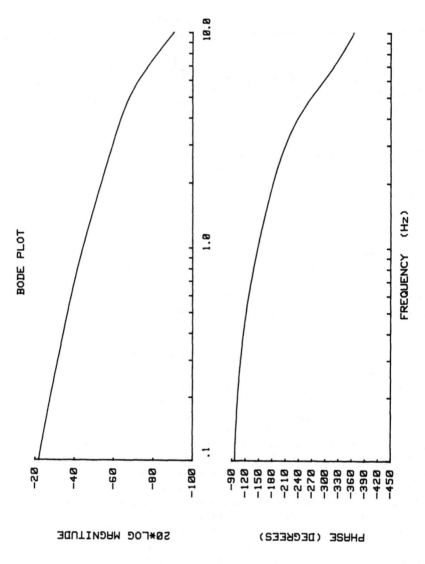

FIGURE 6.25 Minimum plant controller requirements. These are the elements essential for operation prior to loop closure and compensation. The integrator is needed

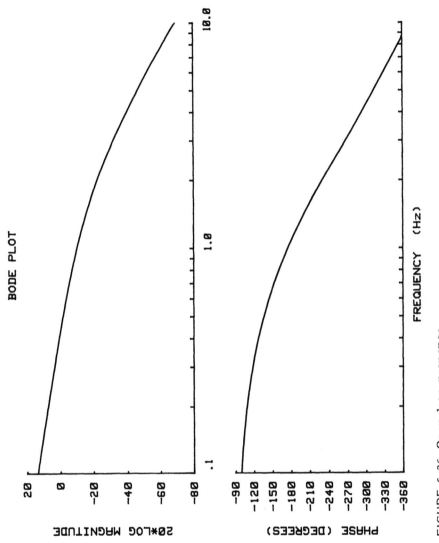

FIGURE 6.26 Open-loop response.

ROOT LOCUS PLOT

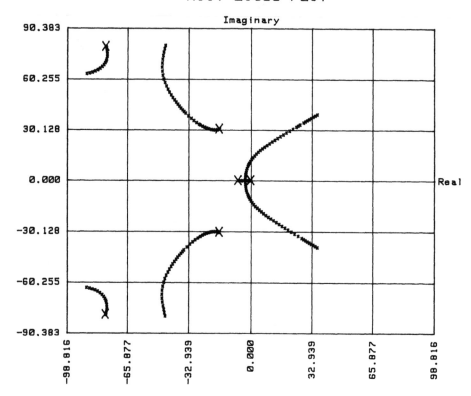

FIGURE 6.27 Root-locus plot of open loop with the motor at maximum displacement.

Figure 6.29 represents the closed-loop dynamics of the system
with K_X varied from 0 to 0.25 (this value corresponds to the first-
order lead time constant). Increasing K_X decreased the amplitude
peak and increased the position (in terms of frequency) of the 90°
phase lag. The amplitude rolloff occurs sooner than desired and
requires more compensation. Figure 6.30 results from adding a
derivative constant (T_d) of 0.025 and changing K_X to 0.1 to fit
the second-order lead damping ratio requirements. Recalling the
second-order interpretation of T_i, T_d, and K_X in terms of its
damping ratio (ζ) and natural frequency (F_n),

ROOT LOCUS PLOT

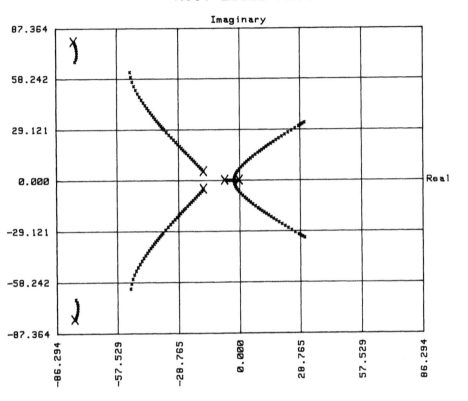

FIGURE 6.28   Root-locus plot with minimum motor displacement.

$$\zeta = \frac{K_x}{2} \frac{1}{\sqrt{T_i/T_d}} \, , \quad F_n = \frac{1}{2\pi} \frac{1}{\sqrt{T_iT_d}}$$

and recalling that T_i has a value of unity, we can see that the natural frequency (F_n) is a function of the derivative time (T_d). This frequency allows for easier placement of the second-order lead, since it represents the bandwidth needed to match to the controlled items (this would be similar to the time constant, if it were a first-order lead). With this value of T_d, the damping

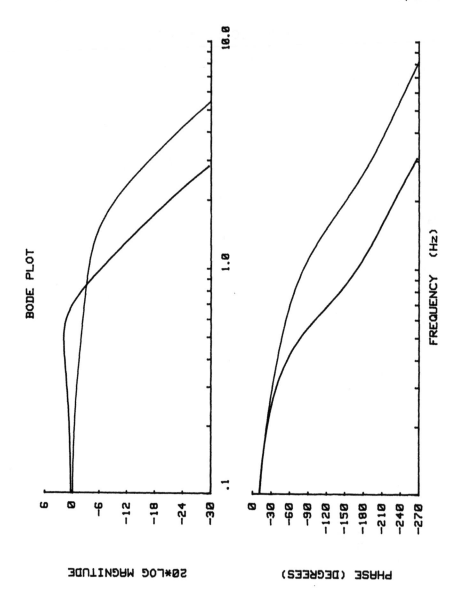

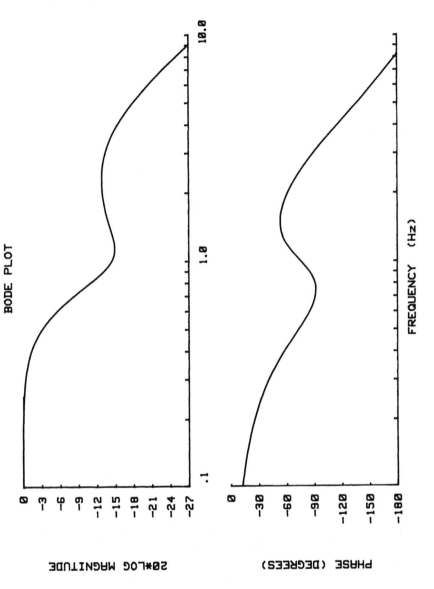

FIGURE 6.30 Inclusion of derivative time constant to the system.

294

Chapter 6

ratio can be set by K_x to implement matching the system to the controller's natural frequency.

Note that Figure 6.30 has good rolloff in both magnitude and phase, up to 1 Hz. Beyond this point both increase due to the increasing compensation of the second-order lead. Since this cannot physically occur, due to limits in the plant, the added phase and magnitude from the controller amplifies noise levels. Additional filtering must occur to account for this region. Such a filter would be similar to a lead-lag circuit, as discussed in Chapter 5.

If the derivative constant were not used, a first-order lead in the feedback (Figure 6.31) could be more beneficial, since it has benefits similar to a lead-lag circuit (because it is in the feedback path). Placement of lead in the feedback acts like lead in the forward path, with the additional benefit of adding a filtering effect after the loop is closed. Figure 6.32 is a closed-loop response of the system with the first-order lead (T_{lead}) of 0.25. The feedback lead has advantages compared to amplifying effects of the derivative-second-order-lead approach. The pseudo-derivative feedback scheme originates from a lead element in the feedback, with system enhancements described in Chapter 5.

6.3.6 Pseudo-Derivative Feedback Speed Loop

The plant in a pseudo-derivative feedback loop can be optimized, first for the inner loop and then the total loop. The pseudo-derivative feedback system of Figure 6.33 has easier implementation of gain settings, due to the placement of the controller gains within the loop. Figure 6.34 shows the inner-loop dynamics of the pump-motor plant discussed previously, with K_{d1} set to 30. It is sized for the appropriate phase and gain margins. Its closed loop (inner loop only) is shown in Figure 6.35

As with the PID loop, an integrator is necessary to maintain zero offset in the output. With the integrator present and its gain (K_i) set to a value of 100, the dynamics of the open outer loop become that of Figure 6.36. The closed loop is shown in Figure 6.37. Since the magnitude rolloff is undesirable, K_{d2} can be utilized; this results in the desirable dynamic response of Figure 6.38. The pseudo-derivative feedback scheme has advantages of simpler implementation and optimization, relative to the PID algorithm.

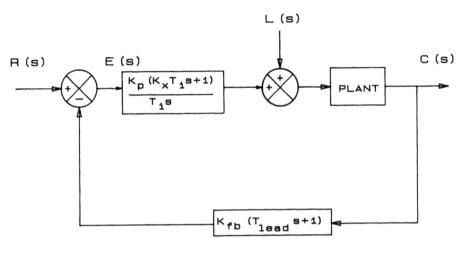

FIGURE 6.31 PI plus lead feedback controller driving pump-motor speed.

6.4 DIGITAL ELECTROHYDRAULIC SYSTEMS

Figure 6.39 is a block diagram of a rotary actuator within a digitally controlled loop. The loop requirements are typically to maintain an output angular position proportional to the loop's input command. It may be used to position a large rotary valve, platform, or test apparatus. With techniques similar to [3], the microcontroller operation for angular velocity feedback can be modified to position feedback. Therefore, the microcontroller combined with the digital Z-transform analysis of Chapter 3, can produce a digital closed-loop system.

The digital filter requirements of the controller, as discussed in Chapter 5 (through the W-plane), will be used to size the filter to the plant and system requirements. The plant is shown with two dominant first-order lag elements. It is kept simple to keep the emphasis in matching the plant to the filter. The time constant T_1 could represent the dominant bandwidth of either a feedback-wire style or the two-spool version of a flow-control servovalve. The time constant T_2 could be the result of an inertially dominant loading situation (see Figure 6.19).

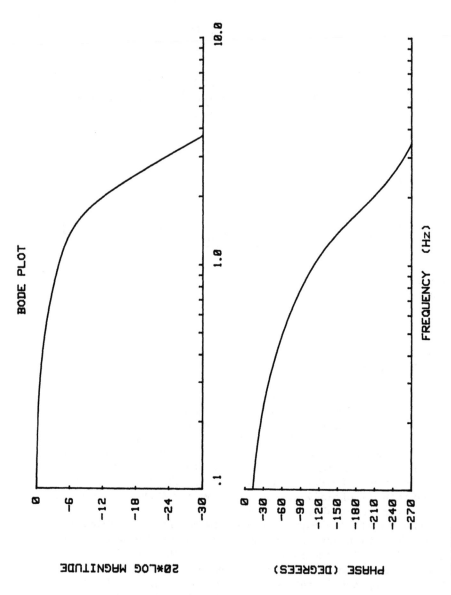

FIGURE 6.32 Closed-loop response effects by adding the first-order lead in the feedback path.

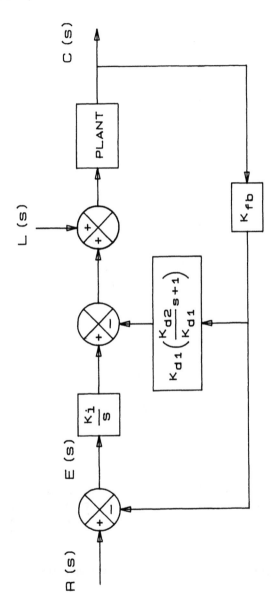

FIGURE 6.33 Pseudo-derivative feedback controlled speed loop.

298

Chapter 6

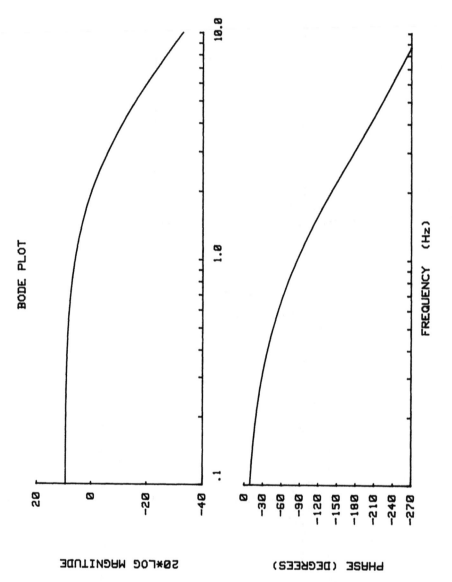

FIGURE 6.34 Pseudo-derivative inner-loop dynamics shown in open-loop form.

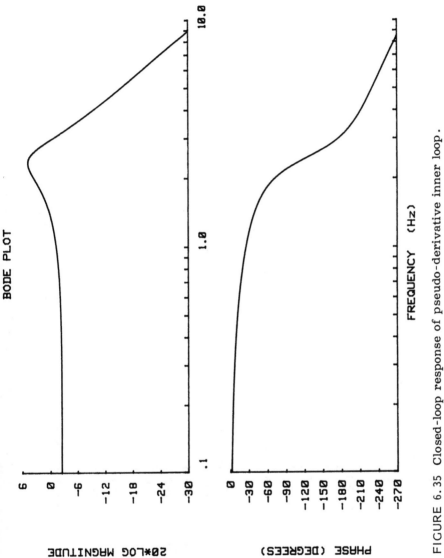

FIGURE 6.35 Closed-loop response of pseudo-derivative inner loop.

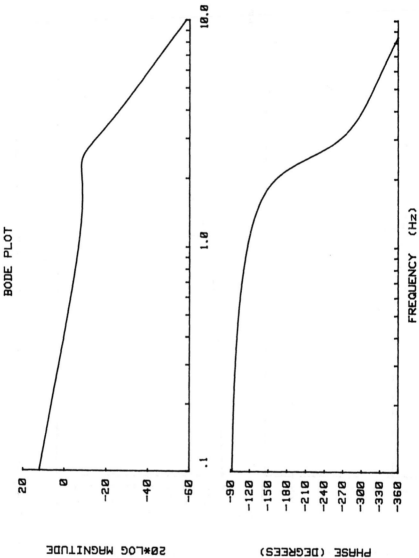

FIGURE 6.36 Open loop of total pseudo-derivative control system.

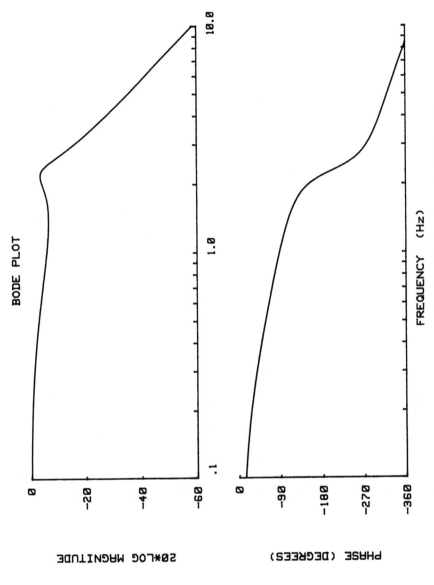

FIGURE 6.37 Closed loop of pseudo-derivative feedback controlled system.

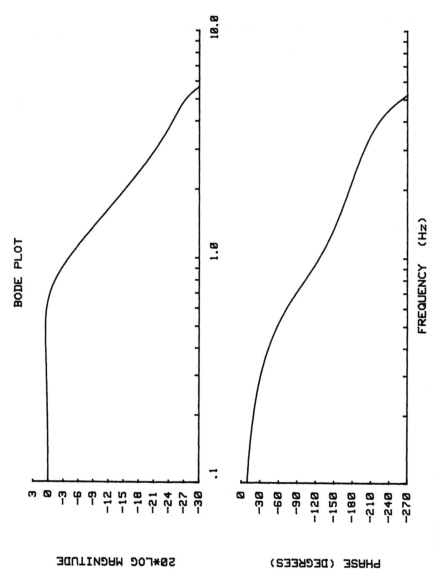

FIGURE 6.38 Closed loop pseudo-derivative feedback enhanced with Kd_2.

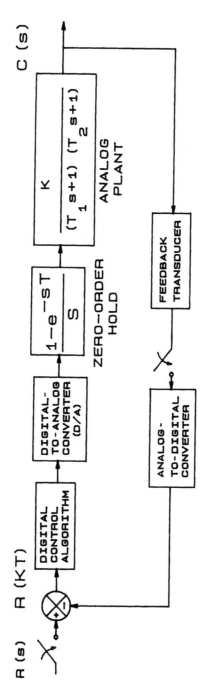

FIGURE 6.39 Digital controller with analog plant.

303

The Z-domain becomes the digital equivalent for directly establishing the controller's characteristics and compensation settings. The bilinear transformation will be used for analyzing
the Z-domain equivalent of the s-domain transfer function. For
high-order systems, this transformation becomes cumbersome,
especially for iterations in obtaining maximized controller settings.

6.4.1 Transformation Implementation

Before proceeding into the loop analysis, we recall from Chapter
3 that the bilinear transform allows us to relate the s-domain to
the Z-domain:

$$\frac{1}{s} = \frac{T_s}{2} \frac{z+1}{z-1}$$

For high-order systems, this transformation would become very
tedious. Note also, from Chapter 5, that the Z- and W-planes
are related by

$$W = \frac{z-1}{z+1}$$

This transformation would become equally cumbersome. If the
bilinear transform is inverted (to use s rather than 1/s), the
equation becomes

$$s = \frac{2}{T_s} \frac{z-1}{z+1}$$

Except for the factor $1/T_s$, this has the same form as the W transform. A very convenient method has been established [4], which
minimizes the implementation of the transforms. The goal of the
algorithm is to progress from a polynomial containing a function
of X or P(x) to a polynomial P[(x - 1)/(x + 1)]. The following
steps, when implemented for the nth-order s-domain transfer
function, will produce a very effective approach in transforming
the s- to Z-domains through the bilinear transformation:

1. P(x) into P(x + 1)
2. P(x + 1) into P(1/x + 1)

3. $P(1/x + 1)$ into $P(-2/x + 1)$
4. $P(-2/x + 1)$ into $P[-2/(x + 1) + 1]$
5. $P[(x - 1)/(x + 1)] = P[-2/(x + 1) + 1]$

To obtain a feel for $P(x + 1)$, consider the second-order polynomial

$$P(x) = a_0 + a_1 x + a_2 x^2$$

If this is expanded to $P(x + 1)$, the result is

$$P(x + 1) = a_0 + a_1(x + 1) + a_2(x + 1)^2$$

$$= (a_0 + a_1 + a_2) + (a_1 + 2a_2)x + a_2 x^2$$

The effect is to make the following replacements:

$$a_0 + a_1 + a_2 \rightarrow a_0, \quad a_1 + 2a_2 \rightarrow a_1, \quad a_0 \rightarrow a_2$$

The polynomial $P(x + 1)$ can be generalized [5] for an nth-order system; the replacement coefficients are then given by

$$a_i = \sum_{k=i}^{n} \frac{k!}{(k - i)! i!} a_k$$

Transfer functions in the s-domain can be expressed as the ratio of two polynomials, according to

$$G(s) = \frac{\sum_{i=0}^{m} a_i s^i}{\sum_{i=0}^{n} b_i s^i} = \frac{N(s)}{D(s)} \quad \text{where } n \geq m$$

Similarly, the Z-domain analysis contains polynomials in Z. Note that the algorithm works independently for the numerator and denominator. Since the s-to-Z transfer involves the sampling time, the algorithm will vary according to the sampling time section. The general form to be implemented by the algorithm is

$$G_d(z) = \sum_{i=0}^{m} a_i \left(\frac{2}{T_s}\right)^i \left[\frac{z-1}{z+1}\right]^i \Big/ \sum_{i=0}^{n} b_i \left(\frac{2}{T_s}\right)^i \left[\frac{z-1}{z+1}\right]^i \quad \text{where } n \geq m$$

where the factor $2/T_s$ is ignored for the inverse W transform, and a_i is replaced by $a_i(2/T_s)^i$.

The procedure for implementing the algorithm is as follows:

1. Replace each a_i with

$$\sum_{k=i}^{n} \frac{k!}{(k-i)!\,i!}\, a_k$$

2. Reverse the order of coefficients wherein
    a_0 is replaced by a_n.
    a_1 is replaced by a_{n-1}.
    a_i is replaced by a_{n-i}.
3. Multiply the coefficients by $(-2)^{n-i}$.
4. Repeat 1: replace each a_i with the binomial expansion.
5. The result is the polynomial expansion.

For example, for the analog transfer function

$$G(s) = \frac{1}{s^2 + 2s + 1}$$

the Z-domain transfer function is (with a sampling time of 1 s)

$$G(z) = \frac{z + 2z + 1}{9z^2 - 6z + 1}$$

Numerator $s^0 = 1$

		a_2	a_1	a_0
1.	a_i (original set)	0	0	1
2.	$a_i(2/T_s)^i$	0	0	1
3.	$a_2 = a_2$	0	0	1
	$a_1 = 2a_2 + a_1$			
	$a_0 = a_0 + a_1 + a_2$			
4.	Reverse coefficients	1	0	0
5.	$a_i(-2)^{n-i}$	1	0	0
6.	Repeat 3	1	2	1

Denominator $s^2 + 2s + 1$

	a_2	a_1	a_0
1. a_i (original set)	1	2	1
2. $a_i(2/T_s)^i$	4	4	1
3. $a_2 = a_2$	4	12	9
$a_1 = 2a_2 + a_1$			
$a_0 = a_0 + a_1 + a_2$			
4. Reverse coefficients	9	12	4
5. $a_i(-2)^{n-i}$	9	-24	16
6. Repeat 3	9	-6	1

The following BASIC program combines the algorithm steps for the transform from s-plane to Z-plane.

```
10   REM        S-TO-Z BILINEAR TRANSFORM
20   REM DIMENSION (M,I,J) M=0 ----- A COEFFICIENT
30   REM                   =1 ----- A COEFFICIENT-temporary
40   REM                   I  ----- Ai COEFFICIENT
50   REM                   J=0 ----- NUMERATOR
60   REM                   =1 ----- DENOMINATOR
70   INPUT "INPUT THE SAMPLING TIME CONSTANT (Ts)",Ts
80   INPUT "INPUT THE ORDER OF THE DENOMINATOR",N
90   FOR J=0 TO 1
100    IF J=0 THEN
110      PRINT "INPUT NUMERATOR COEFFICIENTS"
120    ELSE
130      PRINT "INPUT DENOMINATOR COEFFICIENTS"
140    END IF
150    FOR I=0 TO N
160      PRINT "i=";I
170      INPUT "INPUT Ai",A(0,I,J)
180    NEXT I
190 NEXT J
200 REM J=0 (NUMERATOR)   J=1 (DENOMINATOR)
210 FOR J=0 TO 1
220    FOR I=0 TO N
230    A(0,I,J) = A(0,I,J)*(2/Ts)^I
240    NEXT I
250    REM  OBTAIN P(x+1):
260    GOSUB Xplus1
```

```
270     REM  REVERSE COEFFICIENTS:
280     FOR I=0 TO N
290       A(1,I,J) = A(0,N-I,J)
300     NEXT I
310     FOR I=0 TO N
320       A(0,I,J) = A(1,I,J)
330     NEXT I
340     REM SCALE Ai:
350     FOR I=0 TO N
360       A(0,I,J) = A(0,I,J)*(-2)^(N-I)
370     NEXT I
380     REM  OBTAIN P(x+1):
390     GOSUB Xplus1
400 NEXT J
410 FOR J=0 TO 1
420   IF J=0 THEN
430     PRINT "NUMERATOR COEFFICIENTS:"
440   ELSE
450     PRINT "DENOMINATOR COEFFICIENTS:"
460   END IF
470   FOR I=0 TO N
480     PRINT "A";I;"=";A(0,I,J)
490   NEXT I
500 NEXT J
510 Xplus1: REM SUMMATION ROUTINE
520 FOR I=0 TO N
530   V=0
540   FOR K=I TO N
550     V=V+FNFac((K))/(FNFac((K-I))*FNFac((I)))*A(0,K,J)
560   NEXT K
570   A(1,I,J)=V
580 NEXT I
590 REM  STORE NEWEST Ai
600 FOR I=0 TO N
610   A(0,I,J)=A(1,I,J)
620 NEXT I
630 RETURN
640 END
650 DEF FNFac(M)
660   IF M=0 THEN RETURN 1
670   RETURN M*FNFac(M-1)
680 FNEND
```

6.4.2 Digital Speed Loop

The linearized, closed-loop analog frequency response is shown in Figure 6.40. By changing the controller to a digital controller, the analysis requirements involve transformation into the Z-domain; this includes the hold-device dynamics. Since the zero-order-hold connects to an analog device, it is a part of $GH*(s)$ (refer to the discussion of the pulse transfer function in Chapter 3). This is needed in order to size the digital filter (compensator).

The first step in the analysis is to convert the plant (including the zero-order hold, since it is effectively part of the plant) into the Z-domain. Recall from Chapter 3 that the transformation is

$$G(z) = Z[\text{zero-order hold}]\,Z[\text{plant}]$$

$$G(z) = Z\left[\frac{1 - e^{-T_s s}}{s}\right] Z[G(s)]$$

$$= Z\left[1 - e^{-T_s s}\right] Z\left[\frac{G(s)}{s}\right] = (1 - z^{-1})Z\left[\frac{G(s)}{s}\right]$$

Therefore, for the plant of Figure 6.39,

$$G(z) = (1 - z^{-1})Z\left[\frac{1}{s}\frac{K_o}{s(T_1 s + 1)(T_2 s + 1)}\right]$$

$$= \frac{z - 1}{z} Z\left[\frac{T_1 T_2 K_o}{s(s + 1/T_1)(s + 1/T_2)}\right]$$

$$= \frac{z}{z - 1} Z\left[\frac{2000K_o}{s(s + 100)(s + 20)}\right]$$

Partial-fraction expansion results in

$$G(z) = \frac{z}{z - 1} Z\left[\frac{1}{s^2} - \frac{0.06}{s} - \frac{0.0025}{s + 100} + \frac{0.0625}{s + 20}\right]K_o$$

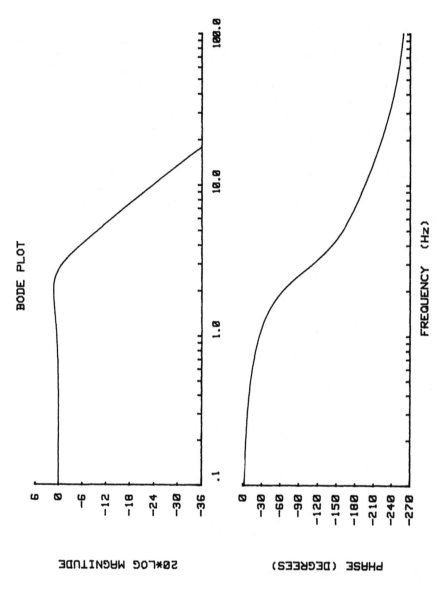

FIGURE 6.40 Analog open-loop frequency-response plot of Figure 6.39.

From the table of Appendix 3, the Z-domain equivalent of the plant becomes (wherein a sampling period of $T_S = 0.05$ s has been used):

$$G(z) = \frac{z - 1}{z} \left(\frac{T_s z}{(z - 1)^2} - \frac{0.06z}{z - 1} - \frac{0.0025z}{z - e^{-100T_s}} + \frac{0.0625z}{z - e^{-20T_s}} \right) K_o$$

$$= \left(\frac{T_s}{z - 1} - 0.06 - \frac{0.0025(z - 1)}{z - e^{-100T_s}} + \frac{0.0625(z - 1)}{z - e^{-20T_s}} \right) K_o$$

$$G(z) = \frac{0.128z^2 + 0.182z + 0.01}{z^3 - 1.37z^2 + 0.368z} K_o$$

In order to size the digital filter, the open-loop system $G(z)$ must be transformed into the W-plane. The computer program for the bilinear transformation assumed the form

$$s = \frac{2}{T_s} \frac{z - 1}{z + 1}$$

The W transform $W = (z - 1)/(z + 1)$ must be rearranged to obtain W from Z or $z = (1 + W)/(1 - W)$, which is effectively (disregarding the sampling period T_S) the negative inverse of the bilinear equation. The transformation can be accomplished simply by altering the Z-transform representation of the system and (slightly) modifying the bilinear transformation computer program. The inverse requirement can be handled by obtaining the inverse of the Z-domain expression. Divide the numerator and denominator by Z^n (where n is the highest order of the polynomial):

$$G(z) = \frac{z^{-3} \{0.128^2 z + 0.182z + 0.01\}}{z^{-3} \{z^3 - 1.37z^2 + 0.368z\}}$$

$$= \frac{0.128z^{-1} + 0.182z^{-2} + 0.01z^{-3}}{1 - 1.37z^{-1} + 0.368z^{-2}}$$

The modified algorithm is accomplished according to the following procedure:

1. $P(x)$ into $P(x - 1)$
2. $P(x - 1)$ into $P(1/x - 1)$

3. P(1/x - 1) into P(2/x - 1)
4. P(2/x - 1) into P[2/(x + 1) - 1]
5. P[2/(x + 1) - 1] = P[(1 - x)/(1 + x)]

The steps which vary from the bilinear transformation are 3 and
5. Therefore 5 is handled easily by excluding the negative sign,
and 3 is accomplished by changing the binomial expansion to

$$\sum_{k=i}^{n} \frac{k!}{(k - i)!i!} \, a_k(-1)^{k-i}$$

In the computer program, lines 250, 260, and 360 change to

250 REM OBTAIN P(x-1)
260 GOSUB Xminus1
270 A(0,I,J)=A(0,I,J)*2^(N-I)

Subroutine Xminus1 is equivalent to Xplus1, with the addition of
the sign in line 1550:

1510 REM Xminus1: MODIFIED SUMMATION ROUTINE
1550 V=V+FNFac((K))/(FNFac((K-I))*FNFac((I)))*A(0,K,J)*
      (-1)^(K-I)

When the coefficients are entered into the computer program,
each a_i is the coefficient of Z^{-i}. The W transform, utilizing the
modified form of the bilinear algorithm, yields

$$G(W) = \frac{0.044W^3 - 0.28W^2 - 0.084W + 0.32}{2.17W^3 + 3.17W^2 + W} K_o$$

$$= \frac{0.138W^3 - 0.875W^2 - 0.263W + 1}{2.17W^3 + 3.17W^2 + W} (0.25K_o)$$

By utilizing the W-plane analysis of Chapter 3 and the bilinear-
transform algorithm, we can investigate the plant's requirements
for closed-loop control. After conclusions have been drawn (the
compensation requirements), the W transform of the digital filter
will be multiplied by the existing W polynomials to obtain the desired frequency response for stability (on the W-plane). Once
these parameters have been created, the digital controller's variables can be established.

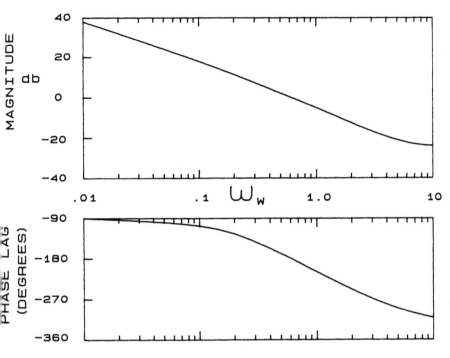

FIGURE 6.41 W-plane frequency response of digital form of Figure 6.39.

The W-plane plot of the open-loop system is shown in Figure 6.41 with a gain of 3, a gain margin of 0 dB, and a phase margin of 0°. In order to obtain sufficient phase and gain margins, we add a lag circuit. The lag circuit, as discussed in Chapter 5, has the form (analogous to the s-domain)

$$G_c(W_\omega) = \frac{\alpha T W_\omega + 1}{T W_\omega + 1} = \text{W-domain lag circuit} \quad \text{where } \alpha < 1$$

With T = 2000 and α = 0.25, the lag circuit (shown dynamically in Figure 6.42) alters the open loop dynamics to Figure 6.43. The phase and gain margins are improved to 45° and 12 dB, respectively. Recall from Chapter 3 that the Bode frequency-

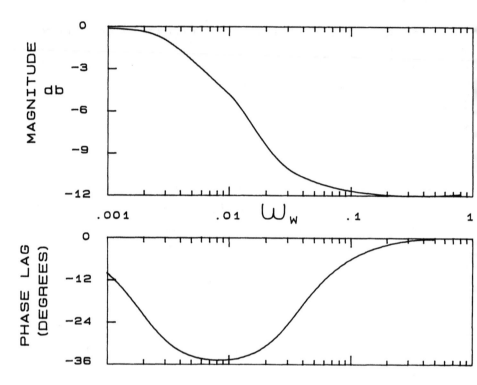

FIGURE 6.42 Frequency response of digital filter for Figure 6.41.

response plots and the W-plane-response plots are related by $W_\omega = \tan(\omega T_S/2)$. This can be simplified, when $\omega T_S/2 \ll 1$, to

$$W_\omega = \frac{\omega T_S}{2}$$

Therefore, with a sampling period of 0.05 s and a W_ω frequency of 0.5 rad/s for the open-loop frequency response on the W_ω-plane, the equivalent frequency in the continuous plane is

$$\omega = \frac{2W_\omega}{T_S} = 2\frac{0.5}{0.05} = 20 \text{ rad/s or } 3.2 \text{ Hz}$$

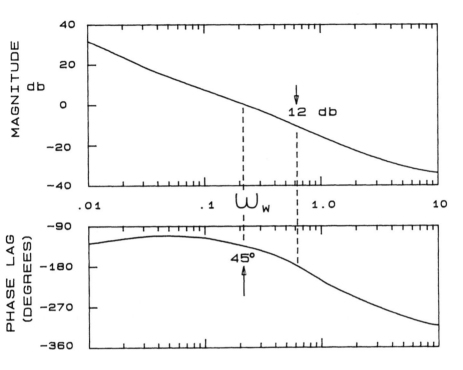

FIGURE 6.43 Compensated open-loop dynamics.

The Z-domain equivalent, using the bilinear transformation algorithm (without including the sampling time), gives

$$\text{lag filter} = \frac{z - 0.96}{z - 0.99} \quad (0.25)$$

This W-plane analysis is similar to [6] and [7]. The algorithm for this lag circuit was presented in Section 5.7. Figure 6.44 is one approach to implementing the lag circuit (discussed in Chapters 3 and 5). Note that the actual numbers are extended to the block to keep the digital filter algorithm a separate entity.

The complete loop results from the analog plant and the zero-order hold, the digital lag filter, and the feedback position transducer. The A/D conversion of the system is similar to the discussion in Appendix 2. The closed loop forms a digital system.

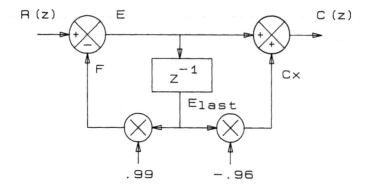

FIGURE 6.44 Lag circuit implementation for the microprocessor.

One could obtain a digital time response through the tedious procedure from Section 3.8.4. The state-space approach to systems
will be shown to be a good, less tedious method in obtaining time
response for analog or digital systems.

6.5 NONLINEAR ANALYSIS THROUGH
MODERN CONTROL THEORY

The advantages of linearized systems analysis, combined with suitable computer programs, are short turn-around time and flexibility between time and frequency responses. Development time is
short because the computer programs (Appendix 1) are set up
with prompts that fit the control loop(s) to the application. There
are many advantages to being able to analyze the transfer function in either the time or frequency domain; each method has its
benefits. The time response is usually the response of interest
to the end user of a control package. Frequency response is
easier to utilize in the development stages of design, stabilization,
and maximizing performance (especially when plant variations are
experimentally determined in the frequency domain by use of the
FFT analyzers).
    Nonlinearities are common in electrohydraulic systems, from
electronic saturations to square-root flow equations. Although
linearization methods are adequate for many systems over a

typical operating range, the analysis utilized should include nonlinear techniques if the variance over wide operating ranges is
of major concern. Several software packages are available for
studying nonlinearities in control systems. A technique utilizing
state-space methods will be presented to analyze nonlinear control loops. In order to use this software technique, we next discuss the basic theory of state-space engineering.

Modern control theory draws heavily from state-space analysis.
It is coined "state-space analysis" because the approach minimizes
a control system to its smallest combination of n variables, called
"state variables," which completely describe the system. The
state variables are analyzed in an n-dimensional space, or state-
space, of which the state variables are represented as elements
of vectors.

This technique expands the investigation of control loops to
systems which have more than one input and output. The mathematical analysis, from simple to complex systems, is based on the
system equations reducing to a form of first-order differential
equations. We replace m nth-order differential equations by n + m
first-order equations. The Laplace representation of this system
of equations can be treated with matrix techniques. The matrix
form is very useful for stability investigation, compensation techniques, and simplicity of analysis.

6.5.1 State-Space Representation

The state-space representation of a control loop is accomplished
by rewriting the system equations in the form of coupled first-
order differential equations. Consider the proportional-plus-in-
tegral controller driving a speed-control system. The block diagram is shown in Figure 6.45. An equivalent block diagram is
shown in Figure 6.46; this form is useful for analysis in state-
space. Each block is converted into a time equivalent and set up
in terms of a first-order equation. The first-order lag $1/(T_2 s + 1)$
becomes

$$X_1 = \frac{1}{T_2 s + 1} X_2, \qquad X_1 T_2 s + X_1 = X_2$$

This is represented in the time domain by

$$T_2 \dot{X}_1 + X_1 = X_2, \qquad \dot{X}_1 = -\frac{1}{T_2} X_1 + \frac{1}{T_2} X_2$$

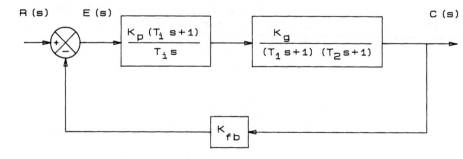

R (s)   E (s)   C (s)

FIGURE 6.45 Proportional plus integral controller controlling
second-order plant.

Similarly, $1/(T_1s + 1)$ is found from

$$X_2 = \frac{K_g}{T_1s + 1} \{X_3 + R(s) - C(s)K_{fb}\}$$

$$\frac{T_1}{K_g} X_2s + \frac{X_2}{K_g} = X_3 + R(s) - K_{fb}X_1$$

$$\frac{T_1}{K_g} \dot{X}_2 + \frac{X_2}{K_g} = X_3 - K_{fb}X_1 + r(t)$$

$$\dot{X}_2 = \frac{-K_{fb}}{T_1} X_1 - \frac{1}{K_g} X_2 - X_3 + r(t)$$

Note that $(T_is + 1)/T_is$ factors as $1 + 1/T_is$; this gives

$$X_3 = (R(s) - K_{fb}X_1) \frac{K_p}{T_is}$$

$$\frac{T_i}{K_p} X_3s = R(s) - K_{fb}X_1$$

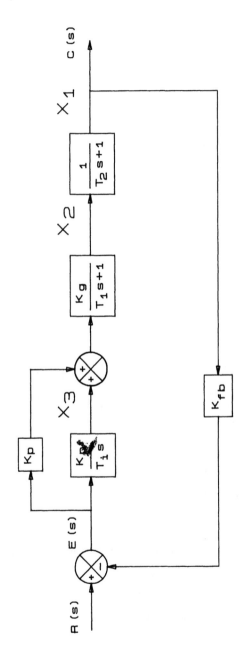

FIGURE 6.46 State representation of Figure 6.45.

$$\frac{T_i}{K_p} \dot{X}_3 = r(t) - K_{fb}X_1$$

$$\dot{X}_3 = \frac{-K_{fb}K_p}{T_i}X_1 + \frac{K_p}{T_i}r(t)$$

Rearranging in terms of $\dot{X}_i$, we obtain the following first-order differential equations:

$$\dot{X}_1 = -\frac{1}{T_2}X_1 - \frac{1}{T_3}X_2$$

$$\dot{X}_2 = -\frac{K_{fb}K_g}{T_1}X_1 - \frac{1}{T_1}X_2 - \frac{K_g}{T_1}X_3 + \frac{K_g}{T_1}r(t)$$

$$\dot{X}_3 = -\frac{K_{fb}K_p}{T_i}X_1 + \frac{K_p}{T_i}r(t)$$

These three equations can be represented in matrix form as

$$\begin{bmatrix} \dot{X}_1 \\ \dot{X}_2 \\ \dot{X}_3 \end{bmatrix} = \begin{bmatrix} -1/T_2 & -1/T_2 & 0 \\ -K_{fb}K_g/T_1 & -1/T_1 & -K_g/T_1 \\ -K_{fb}K_p/T_i & 0 & 0 \end{bmatrix} \begin{bmatrix} X_1 \\ X_2 \\ X_3 \end{bmatrix} + \begin{bmatrix} 0 \\ K_g/T_1 \\ K_p/T_i \end{bmatrix} [r(t)]$$

$$c(t) = \begin{bmatrix} 1 & 0 & 0 \end{bmatrix} \begin{bmatrix} X_1 \\ X_2 \\ X_3 \end{bmatrix}$$

This reduces to a standard form for state-space analysis (shown in Figure 6.47):

$$\dot{X} = AX + Br = \text{the state equation}$$

$$y = CX + Dr = \text{the output vector}$$

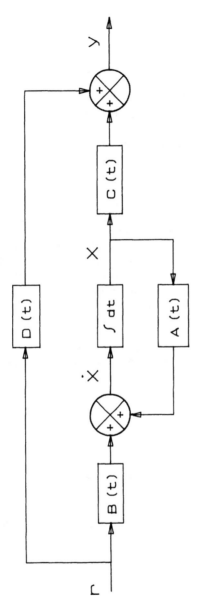

FIGURE 6.47 General state block diagram.

where

A, B, C, D = constant matrices

X, Y   = variable matrices

X   = state vector of the system or component

y   = output vector of the system or component

In this example, the output y is equal to x_1, as shown in the block diagram. In order to solve these equations and therefore determine the output of the system, we must solve the linear matrix equation first for x_n and then evaluate the output y_n. To solve the first-order matrix equation, we give a general form of the matrix representation, with analogies to scalar functions.

Figure 6.47 represents a generalized matrix-style control system. The inner portion of the integrator (with its output x with matrix A in its feedback) basically represents the state equation; each equation was put into the form of a first-order differential equation, "revolving" around an integrator. B represents the matrix of terms involved in the input which directly affect the state equation of the system. D is the matrix representation of the influence of the input acting on the output.

Control theory representation of a generalized system results in the following state variables for a system of n variables. Note that there are derivatives of the forcing (input) as well as of the state variables.

$$X_1 = y - \beta_0 r$$

$$X_2 = (\dot{y} - \beta_0 \dot{r}) - \beta_1 r = \dot{x}_1 - \beta_1 r$$

$$X_3 = (\ddot{y} - \beta_0 \ddot{r} - \beta_1 r) - \beta_2 r = x_2 - \beta_2 r$$

$$x_n = \left(\overset{n-1}{y} - \beta_0 \overset{n-1}{r} - \beta_1 \overset{n-2}{r} \right) - \cdots - \beta_{n-2} \dot{r} - \beta_{n-1} r = \dot{x}_{n-1} - \beta_{n-1} r$$

where

$$\beta_0 = b_0$$

$$\beta_1 = b_1 - a_1 \beta_0$$

$$\beta_2 = b_2 - a_1 \beta_1 - a_2 \beta_0$$

$$\beta_3 = b_3 - a_1\beta_2 - a_2\beta_1 - a_3\beta_0$$

$$\beta_n = b_n - a_1\beta_{n-1} - \cdots - a_{n-1}\beta_1 - a_n\beta_0$$

With these definitions, the generalized state and output equations become

$$
\begin{bmatrix} \dot{x}_1 \\ \dot{x}_2 \\ \dot{x}_{n-1} \\ \dot{x}_n \end{bmatrix} =
\begin{bmatrix} 0 & 1 & 0 & \cdots & 0 \\ 0 & 0 & 1 & \cdots & 0 \\ 0 & 0 & 0 & \cdots & 1 \\ -a_n & -a_{n-1} & & \cdots & -a_1 \end{bmatrix}
\begin{bmatrix} x_1 \\ x_2 \\ x_{n-1} \\ x_n \end{bmatrix} +
\begin{bmatrix} \beta_1 \\ \beta_2 \\ \beta_{n-1} \\ \beta_n \end{bmatrix} [r]
$$

$$
y = [1 \ 0 \ \cdots \ 0]
\begin{bmatrix} x_1 \\ x_2 \\ \cdot \\ \cdot \\ \cdot \\ x_n \end{bmatrix} + \beta_0 r
$$

This is equivalent to the matrix equations

$$\dot{x} = Ax + Br, \quad y = Cx + Dr$$

6.5.2 Solving the State Equation

The state equation can be solved by comparing it to a scalar system such as the first-order equation (with no input)

$$\dot{x} = ax$$

This equation has Laplace transform

$$sX(s) - x(0) = aX(s)$$

$$X(s)[s - a] = x(0)$$

$$X(s) = \frac{x(0)}{s - a} = (s - a)^{-1}x(0)$$

The time-domain solution is thus

$$x(t) = e^{at}x(0)$$

Similarly, this can be extended into matrix form, for a set of coupled first-order equations

$$X(t) = AX(t)$$

The Laplace transform gives

$$sX(s) - x(0) = AX(s), \quad (sI - A)X(s) = x(0)$$

where I is the identity matrix. Since the order of multiplication and inversion must be maintained in matrix algebra, the equation is solved for the matrix X(s) as

$$X(s) = (sI - A)^{-1}x(0)$$

Define $\Phi(s) = (sI - A)^{-1}$.

The time domain becomes

$$\Phi(t) = \mathcal{L}^{-1}[(sI - A)^{-1}]$$

Therefore the zero-input solution becomes

$$X(t) = \Phi(t)x(0)$$

The zero-input equation can be rewritten as

$$\dot{X} = AX$$

$$e^{-At}X = e^{-At}AX$$

$$e^{-At}X - e^{-At}AX = 0$$

which is equivalent to

$$\frac{d}{dt}\left(e^{-At}X\right) = 0$$

which reduces to

$$e^{-At}X = \text{constant}$$

Therefore the zero-input solution is

$$X = e^{At}x(0)$$

By comparing the two solutions, we see that the state transition matrix has the alternative form

$$\Phi(t) = e^{At}$$

The time solution for the zero-input equation becomes

$$x(t) = e^{At}x(0) = \Phi(t)x(0)$$

It is important to note that the matrix e^{At} has meaning only in the context of the (Taylor) series

$$e^{At} = I + At + \frac{(At)^2}{2!} + \cdots$$

Obviously, control systems will have inputs or forcing functions which must be accounted for to obtain the total solution of the state equation. The state equation

$$\dot{x}(t) = Ax(t) + Br(t)$$

transforms into the s-domain as

$$sx(s) - x(0) = Ax(s) + Br(s)$$

$$(sI - A)x(s) = x(0) + Br(s)$$

$$x(s) = (sI - A)^{-1}x(0) + (sI - A)^{-1}Br(s)$$

Let $\phi(s) = (sI - A)^{-1}$ (as before). Then

$$\Phi(t) = \mathcal{L}^{-1}[\Phi(s)] = \mathcal{L}^{-1}[(sI - A)^{-1}]$$

The complete s-domain solution is then

$$X(s) = \Phi(s)x(0) + \Phi(s)BR(s)$$

$$= \text{zero-input solution} + \text{zero-state solution}$$

The time-domain solutions obtained from the convolution theorem (which equates multiplication in the s-domain to convolution in the time domain) are

$$X(t) = \Phi(t)x(0) + \mathcal{L}[\Phi(s)BR(s)]$$

$$= \Phi(t)x(0) + \int_0^t \Phi(t - \tau)Br(\tau)\,d\tau$$

$$= \Phi(t)x(0) + \int_0^t e^{A(t-\tau)}Br(\tau)\,d\tau$$

$$= \phi(t)x(0) + e^{At}B\int_0^t e^{-A\tau}r(\tau)\,d\tau$$

zero-input part   zero-state part

   The classical method has advantages for analyzing linear systems of this type. The state-space analysis requires rearranging or expanding the terms of the block diagram, whereas the classical method has been maximized with the computer programs for minimal setup time. The state-space method expands analysis into multiple-input and outputs and can include nonlinearities. The controller algorithms can be optimized to a set of criteria demanded for the system. Since nonlinearities are prevalent in hydraulic systems, state-space analysis will be utilized in examining control loops with nonlinear elements. By combining the state equation (and solution) within a computer program which characterizes each portion of the loop as a function of time, the total response (including nonlinearities) can be evaluated.

6.5.3 Handling Nonlinearities Using State-Space Analysis

In order to extend the state equation and its solution into a computer simulation, the solution must be converted into a form usable by the computer. After a method of simulation is established, a system will be shown which uses the state-space method to analyze nonlinearities typically found in electrohydraulic control systems. To modify the state equation

$$X(t) = \Phi(t)x(0) + \int_0^t \Phi(t - \tau)Br(\tau) \, d\tau$$

to fit digital simulation, the state transition matrix becomes

$$\Phi(T) = e^{AT} = I + AT + \frac{A^2 T^2}{2!} + \cdots = I + \sum_{j=1}^{n} \frac{A^j T^j}{j!}$$

The matrix for the zero-state solution can be evaluated by defining the matrix

$$\Phi_0(T) = B \int_0^T \Phi(t) \, dt$$

where

$$\Phi(t) = e^{At} = I + At + \frac{A^2 T^2}{2!} + \cdots = 1 + \sum_{j=1}^{n} \frac{A^j T^j}{j!}$$

By substituting the continuous state transition matrix into the integration, the digital state transition matrix of the input becomes

$$\Phi_0(T) = B \int_0^T \left\{ 1 + \sum_{j=1}^{n} \frac{A^j t^j}{j!} \right\} dt$$

$$= \left[T + \sum_{j=1}^{n} \frac{A^j}{j!} \int_0^T t^j \, dt \right] B$$

$$\Phi_0(T) = \left[T + \sum_{j=1}^{n} \frac{A^j}{j!} \frac{t^{j+1}}{j+1} \Big|_0^T \right] B$$

$$= \left[T + \sum_{j=1}^{n} \frac{A^j}{j!} \frac{T^{j+1}}{j+1} \right] B$$

$$= \left[T \left\{ I + \sum_{j=1}^{n} \frac{A^j T^j}{j!(j+1)} \right\} \right] B$$

$$= \left[T \left\{ I + \sum_{j=1}^{n} \frac{A^j T^j}{(j+1)!} \right\} \right] B$$

$$= \left[T \left\{ I + \frac{AT}{2!} + \frac{A^2 T^2}{3!} + \cdots \right\} \right] B$$

These digital state transition matrices are synonymous with the analog state and allow a system to be simulated on a digital computer. Recall the continuous state equation and solution:

State equation

$$\dot{X} = AX + Br = \text{the state equation}$$

$$y = CX + Dr = \text{the output vector}$$

where

A, B, C, D = constant matrices

X, y   = variable matrices

X   = state vector of the system or component

y   = output vector of the system or component

Solution to state equation (continuous)

$$X(t) = \Phi(t)x(0) + \mathcal{L}\{\Phi(s)BR(s)\}$$

$$= \Phi(t)x(0) + \int_0^t \phi(t-\tau)Br(\tau) \, d\tau$$

$$X(t) = \Phi(t)x(0) + \int_0^t e^{A(t-\tau)} Br(\tau) \, d\tau$$

$$= \Phi(t)x(0) + e^{At}B \int_0^t e^{-A\tau} r(\tau) \, d\tau$$

Recalling from Chapter 5 the analogous form between the analog and digital (differential and difference) equations, we obtain the digital solution to the state equation

$$X(\{K + 1\}T) = \Phi(T)X(KT) + \phi_0(T)r(KT)$$

where $X(\{K + 1\}T)$ and $X(KT)$ are the states of system at the K and K + 1 sample times, respectively. The first part of this algorithm gives the solution to the initial, or zero-input-equation, and the second portion gives the zero-state solution. The zero-input term uses the previously calculated complete solutions as "initial values" for the next calculation.

Implementation of the state equation, its solutions, and its digital representation can be comprehended by an example. Figure 6.48 is an electrohydraulic control system in which an electrohydraulic servovalve is used to move a mass under closed-loop control. It can be easily and quickly analyzed using linear techniques, but it may be desirable to check a larger range of the variables which deviate from their linearized values.

The example has two nonlinearities: saturation and square-root functions. The method [3] allows for time simulation for a variety of inputs and disturbances. The nonlinearities in the state-space interpretation are handled as inputs to the state-space equation in an iterative procedure.

This control loop has three main functional elements: spool stroke and its derivatives, valve output flow and its interface, and ram output position with its derivatives. The spool dynamics originate from Newton's laws of motion, according to

$$\Sigma F = m\ddot{X} = -f_1\dot{X} - K_oX + F_i = -f_1\dot{X} - K_oX + K_ci$$

$$\ddot{X} = -\frac{f_1}{m}\dot{X} - \frac{K_o}{m}X + \frac{K_c}{m}i \qquad (1)$$

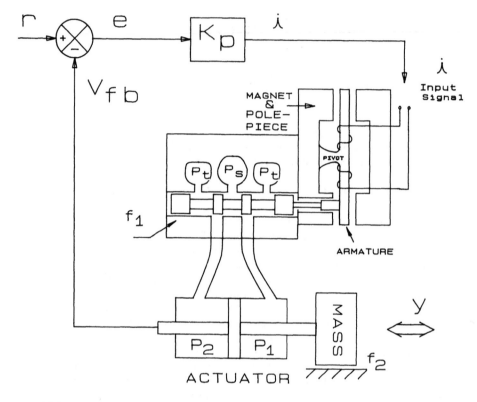

FIGURE 6.48 Position control.

The transformation to the s-domain gives

$$ms^2X(s) + f_1sX(s) + K_OX(s) = F_i(s)$$

where K_O is the equivalent spring rate of the pivot of the pilot.

$$(ms^2 + f_1s + K_O)X(s) = F_i(s)$$

$$X(s) = \frac{1/K_O}{(m/K_O)s^2 + (f_1/K_O)s + 1}F_i(s)$$

$$F_i(s) = K_cI(s) \quad \text{where } I(s) \text{ is the input current}$$

Recall that the basic flow equation is

$$Q = KA_0 \sqrt{\Delta P_v}$$

where

K_0 = fluid flow constant

A_0 = orifice area

ΔP = pressure drop across the orifice

Note also that the formulation of pressure in a chamber or system is

$$P = \int \frac{1}{C_h} Q \, dt, \quad Q = \Sigma q_i$$

where

C_h = equivalent hydraulic capacitance of an element

q_i = individual flow component

After investigating the pressure-flow dependence, we will transform the flow equation to the necessary format. Rewriting the pressure equation in terms of its derivative, we obtain

$$\dot{P} = \Sigma \frac{Q_i}{C_h} = \frac{Q_c - Q_l}{C_h} = \frac{Q_{source} - Q_{load} - Q_l}{C_h}$$

where Q_l = leakage flow.

In terms of the pressures P_1 and P_2,

$$\dot{P}_1 = \frac{(q_1 - A\dot{y}) - K_{li}(P_1 - P_2) - K_{le}P_1}{C_h} \quad (2)$$

$$\dot{P}_2 = \frac{(-q_2 + A\dot{y}) + K_{li}(P_1 - P_2) - K_{le}P_2}{C_h} \quad (3)$$

The flow equations are a function of the spool stroke X and the pressures P_1 and P_2:

$$q_1 = K_q X \sqrt{P_s - P_1}, \quad q_2 = K_q X \sqrt{P_2 - P_t}$$

Note that the variables X, P_1, and P_2 appear in these nonlinear equations as well as in the spool force summation and pressure-flow chamber equations. Combining the equations is not advisable because the resulting compounded nonlinearities are not solvable by the following state-space technique. Instead, the nonlinearities will show up only through the loop iterations as inputs (boundary conditions) for the state equations. Before we set up the state equation and solutions, we will include the dynamic influence of the ram.

Newton's second law applied to the ram yields

$$\Sigma F = M\ddot{y} = -f_2 \dot{y} + F_{i2} \quad \text{where } F_{i2} = (P_1 - P_2)A$$

$$\ddot{y} = -\frac{f_2}{M}\dot{y} + \frac{(P_1 - P_2)A}{M} \qquad (4)$$

Note the equivalent s-domain expression for this dynamic section:

$$(Ms^2 + f_2 s)Y(s) = F_{i2}(s)$$

$$Y(s) = \frac{1/f_2}{s[(M/f_2)s + 1]} F_{i2}(s)$$

Equations (1) through (4) represent four possible state variables, with two more resulting from the integrals of the two second-order equations involving X and Y. By defining the following state variables, equations (1)--(4) can be determined:

$$X_1 = X, \quad X_2 = \dot{X}_1 = \dot{X}, \quad X_3 = P_1,$$

$$X_4 = P_2, \quad X_5 = \dot{X}_6 \ (=\dot{y}), \quad X_6 = Y$$

With these state variables, equations (1)--(4) can be rewritten as first-order equations:

$$\dot{X}_1 = X_2$$

$$\dot{X}_2 = -\frac{f_1}{m}X_2 - \frac{K_o}{m}X_1 + \frac{K_c}{m}i \qquad \{\text{equation (1)}\}$$

$$\dot{X}_3 = \frac{(q_1 - AX_6) - K_{li}(X_3 - X_4) - K_{le}X_3}{C_h} \qquad \{\text{equation (2)}\}$$

$$\dot{X}_4 = \frac{(-q_2 + AX_6) + K_{li}(X_3 - X_4) - K_{le}X_4}{C_h} \qquad \{\text{equation (3)}\}$$

$$\dot{X}_5 = -\frac{f_2}{M}X_6 + \frac{X_3 X_4 A}{M} \qquad \{\text{equation (4)}\}$$

$$X_6 = X_5$$

These equations can be arranged in matrix form:

		X_1	X_2	X_3	X_4	X_5	X_6		
$\ddot{X}$	$\dot{X}_1$	0	1	0	0	0	0		X_1
$\dot{X}_1$	$\dot{X}_2$	$-K_o/m$	$-f_1/m$	0	0	0	0		X_2
$\dot{P}_1$	$\dot{X}_3$	0	0	$-(K_{li}+K_{le})/C_h$	K_{li}/C_h	$-A/C_h$	0		X_3
$\dot{P}_2$	$\dot{X}_4$	0	0	K_{li}/C_h	$-(K_{li}+K_{le})/C_h$	A/C_h	0	=	X_4
$\dot{y}$	$\dot{X}_5$	0	0	A/M	$-A/M$	$-f_2/M$	0		X_5
$\ddot{y}$	$\dot{X}_6$	0	0	0	0	1	0		X_6

	0	0	0	i
	K_c/m	0	0	q_1
+	0	$1/C_h$	0	q_2
	0	0	$-1/C_h$	
	0	0	0	
	0	0	0	

or $X = AX + Br$.

The state variables X, P_1, and P_2 appear in the state equation and the flow equations. The flow equation cannot be included in the state equation (even though X, P_1, and P_2 appear) due to the nonlinearity of the square root. The key to consistent and proper simulation is matching the values of the variables of the solution with the values of the variables in the flow equation through constant updating of loop information. From a set of initial conditions (with spool and ram centered), initial values in the loop can be assumed or calculated. With the spool centered, the starting value of the state variable X is equal to zero.

There are, however, nonzero values for P_1 and P_2. Initially, these can be taken from valve or other component parts or from previous knowledge of the valve settings. Often valves are centered with equal laps at supply and tank, producing a static pressure of one-half the supply pressure at each output port. At each successive iteration, the previous solution becomes the new starting value; it also fixes P_1 and P_2 for that iteration. How then can we obtain accuracy from this division of state variables between the state equation and a secondary nonlinear equation?

By studying the flowchart of Figure 6.49, one can see that the state variables are the initial conditions or results from the last time-sample solution of the state equation up to the point where the "state input" is noted. From this point to the "state solution" is where the state equation is solved. These solutions become the inputs for the next loop (next time sample). The proper time sample becomes the key to the split. The state variables X, P_1, and P_2 are denoted by $X(1, 1, K_t)$, $X(3, 1, K_t)$, and $X(4, 1, K_t)$, respectively; in the flow equations Q_1 and Q_2 values are evaluated at time $K_t T_s$.

The state variables X, P_1, and P_2, which are the solution of the state equation at time $(K_t + 1)T_s$, differ from these same variables at the state input. The proper sizing of these state variables (before and after the state equation is solved) is keyed to the values of the state input (U). Therefore, if Q_1 and Q_2 do not change significantly from time $K_t T_s$ to time $(K_t + 1)T_s$, then the split between the independent formulation of the state variables X, P_1, and P_2 is justified. The object is to keep the change in U for a change in sampled time to an absolute minimum.

The desired sampling time can be derived by two methods: iterations and matching to plant dynamics. Iterations of the entire loop, using smaller and smaller sampling times, will eventually result in solution plots which are identical to plots obtained at a larger sampling time. This would be the adequate sampling time to use, since no more improvement results from further decreases.

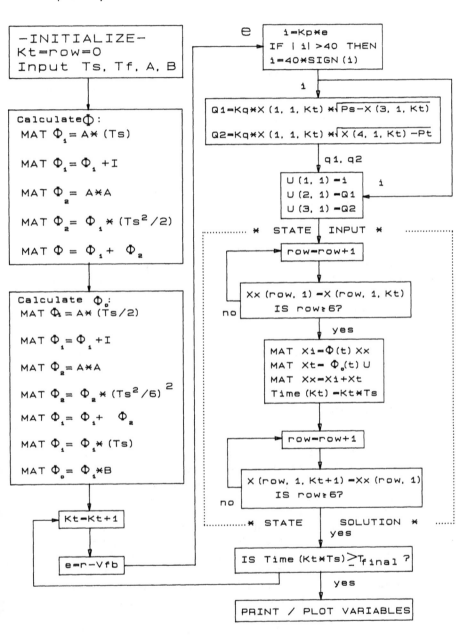

FIGURE 6.49 Nonlinear simulation flowchart for Figure 6.48.

This technique can become quite time consuming, especially since error messages can easily result. The error messages can be an upshot of improper sampling time, wherein the mismatch between $Q_1(K_t*T_S)$ and $Q_1((K_t + 1)*T_S)$ may cause improper levels of P_1 and P_2 (resulting in a square root of a negative number). The value of P_1 may rise too rapidly for large sampling times, where, in reality, the pressure buildup is slower.

Sampling time, as determined by plant dynamics, may also give an acceptable estimate of a proper sampling time. The introduction to Chapter 6 produced a ballpark estimate of the bandwidth of a system, based on the open-loop gain K_0. The result was $F_c = K_0/2\pi$, where F_c is the cutoff frequency (or bandwidth) of the system. As a rule of thumb, if T_S is chosen to be the inverse of 100 times the bandwidth $[T_S = 1/100F_c = 2\pi/100K_0]$, the sampling should be adequate for the loop(s). Fast-acting loading situations and unaccounted for dynamic terms can lead to incorrect solutions; these factors must be accounted for in the determination of T_S as well as in all possible variables. It would be wise to run the model with several time steps to be sure the solutions converge for the chosen T_S.

Figure 6.50 represents the state variables P_1, P_2, Y, and Y' as functions of time for a step input of 5V. Figure 6.51 shows the spool's position and velocity response for the same time span and for a time span indicating saturation.

6.6 CONCLUSION

Whether using proportional values in a closed loop with the human element of feedback or maximizing an adaptive digital controller to a sophisticated closed loop, system sizing is involved. Obviously, the simpler the system, the less demanding is the analysis. Even though control analysis is itself a tool, the more tools used under its implementation to produce stable, responsive loops, the better the resulting system will become.

The more responsive each element of a system becomes, the more responsive will be the total system. Since the ideal system is one which the ideal feedback element sets the pace of the system by dominating the forward-loop dynamics, efforts should be placed into using good feedback elements and designing or using high-response (negligible dynamics) control elements.

The s-domain approach, especially when analyzed through frequency response, is very valuable for establishing controller

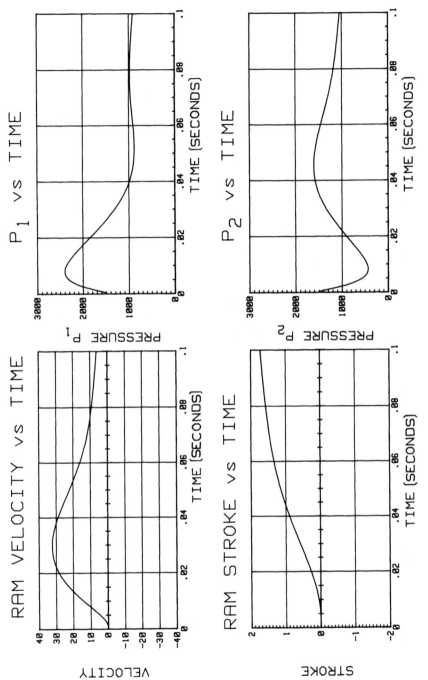

FIGURE 6.50 Ram pressures, position, and velocity dynamics for position control of Figure 6.48.

337

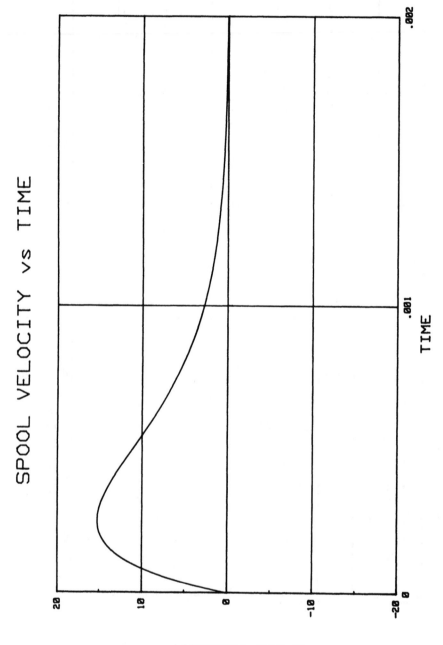

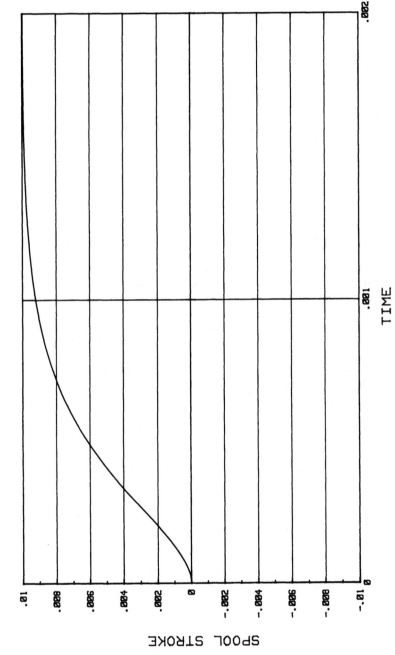

FIGURE 6.51 Spool position and velocity time response of Figure 6.48.

(and digital filter) matchup to plant dynamics. The state-space
analysis becomes powerful when handling multiple-input, multiple-output systems.

The state-space approach, extended to include nonlinearities,
opens up a technique which handles complex, interacting loop elements. The state-space method can also be expanded to handle
digital and mixed A/D systems.

Digital systems have the ability to change with parameter variations. This flexibility is paving the way to making the pumps,
motors, actuators, and valving "smarter" in open- and closed-
loop systems.

BIBLIOGRAPHY

1. Merrit, Herbert E. *Hydraulic Control Systems*, Wiley, 1967.

2. Anderson, Wayne R. System cost and performance advan-
    tages utilizing unique pressure and flow control servovalves,
    SAE Paper 851593, Sept. 1985.

3. Altether, Joseph, and Holm, Robert. New approach to speed
    control, *Machine Design*, Nov. 20, 1986, pp. 173–177.

4. Malvar, Henrique Sarmento. Transform analog filters into
    digital equivalents, *Electronic Design*, April 30, 1981, pp.
    145–148.

5. Anderson, Ronald K. Personal communication, Bemidji State
    University.

6. Kuo, Benjamin C. *Digital Control Systems*, Holt, Rinehart
    and Winston, 1980, pp. 489–493.

7. Intel, notes on the MCS-96 microcontroller chip "Micro-con-
    troller solutions," Intel Corporation.

8. Chen, Han-Sheng. State space analysis of electrohydraulic
    servosystems, *Proceedings of the 1986 Conference of Fluid
    Power*, Vol. 40, 1986, pp. 339--347.

Appendix 1

Computer Programs

Linear analysis of a control system is advantageous in optimizing the system response-stability interface. Nonlinear analysis can be performed with analog, digital, and hybrid (mixed analog and digital) computers. Nonlinear analysis employs time-response analysis; unfortunately, it is more difficult to predict the sizing of compensation elements. The frequency-response method allows one to analyze both open- and closed-loop systems and to optimize the response with suitable compensation and open-loop gains.

The root locus gives some similar insights in assigning the variables of the controller and the compensation. Absolute stability can readily be seen from the maximum gain before crossing into the right half of the s-plane. The location of poles and zeros of the plant determines the need for compensation and identifies the locations for the stabilizing poles and zeros.

Time response for a linear model provides important information, especially when the system is in its fine-tuning stage. Many systems, from machine tool operation to construction equipment, use the time response of their components as their "measuring stick" of acceptance, because the end product must perform within those time constraints. Section 6.5 sets up a good time-domain technique for linear and nonlinear systems.

A general program flowchart will be shown for obtaining the transfer function and performing the root-locus and frequency-

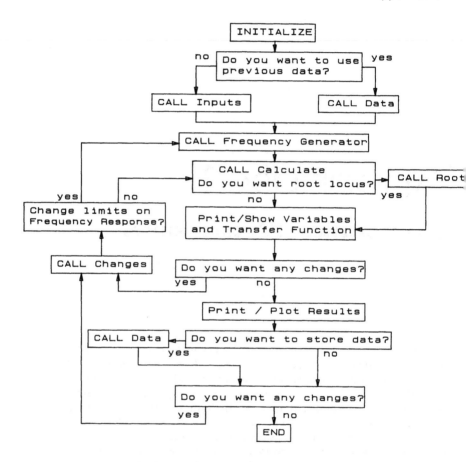

FIGURE A1.1   MAIN flow diagram of dynamic simulation.

response analyses. Flow diagrams for each dynamic method will
be presented. The main flowchart of Figure A1.1 is set up to
employ user-supplied loop parameters for frequency response,
transfer function, and root-locus analyses. This setup allows
input prompts for multiple loop, variable-complexity systems.
Figures A.1.4-A.1.15 give the subroutines CALLed in Fig. A.1.1.
     Once the loop variables are set up in this manner, changing
values becomes easy and efficient. The flowchart of Figure A1.1
portrays the overall scheme in setting up the parameters, calculating the frequency response or root locus (user's choice),
printing the transfer function, plotting the results, and making

Input

Output

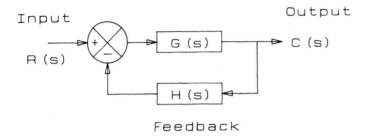

R (s)

G (s)

C (s)

H (s)

Feedback

FIGURE A1.2 General system block diagram.

any changes required for iteration. The (loop) variable changes and plotting routines are not discussed because the methods of implementation vary from computer to computer and with the software employed. The inputs for the loops are tied together with the calculations of the loop elements, since each loop is evaluated as a function of gains, integrators, and first- and second-order leads and lags.

Open and closed loops, as well as loops within loops or side-by-side loops, can be evaluated. The style of the program assumes that the feedback is a unity element. If the loop contains nonunity feedback (Figure A1.2), it can easily be changed to unity feedback (Figure A1.3) by placing the element in the forward loop; after the loop is closed, multiply the loop by the inverse of the feedback element. When the loop parameters are calculated, the user must specify whether root locus or frequency response is desired. If both are desired, the second can be obtained by requesting changes after the first plot is produced.

A general description of the equations used in acquiring the plots will be discussed next. The main inputs are prompts about

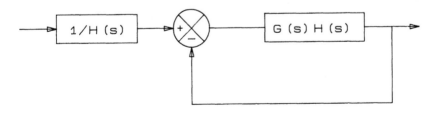

FIGURE A1.3 Unity feedback reduced form of Figure A1.1.

344 Appendix 1

each loop, starting from the innermost loop. For each loop, the choices are the (1) open-loop gain, (2) presence of integrator, (3) number of, and values for, any first-order leads or lags, (4) number of, and values for, any second-order leads and lags, and (5) whether the loop is to be open or closed. When performing the calculations of these parameters, one uses two basic styles.

The method for the frequency response is based on each element (first-order lag, etc.) being evaluated in the frequency domain and then multiplied by the next element. The transfer function and root locus utilize the closed-loop form {T.F. = G(s)/[1 + G(s)H(s)]} in their evaluation. The frequency response could be analyzed by this transform method, but then the transportation lag and experimental results would be omitted.

The transportation lag (time delay in transporting a hydraulic signal over long line lengths) has a corresponding phase lag which varies with frequency. In addition to including the transportation lag, this method can add the results of an actual frequency-response test to the computer-generated dynamics (because the computer program is based on the magnitude and phase of each element in the loop).

The transfer function of a closed-loop system (shown in Figure A1.1) is defined as

$$\text{T.F.} = \frac{G(s)}{1 + G(s)H(s)}$$

Any transfer function which has a nonunity feedback element can be made into a unity feedback equivalent by block-diagram modification. If the feedback element H(s) is changed in position to that of Figure A1.3, an equivalent block diagram with unity feedback exists. The open-loop gain and dynamic response are the same as those indicated by the forward-loop gain and dynamics with transfer function G(s)H(s). The computer program is set up with the assumption that the information about the loop is entered in the form of unity feedback (wherein the user supplies the feedback element as another loop; following the form of Figure A1.3):

$$G'(s) = G(s)H(s)$$

The resulting transfer function of G'(s) becomes

$$\text{T.F.} = \frac{G'(s)}{1 + G'(s)} = \frac{G_n(s)/G_d(s)}{1 + G_n(s)/G_d(s)} = \frac{G_n(s)}{G_d(s) + G_n(s)}$$

where n and d refer to numerator and denominator, respectively. Therefore, the unity feedback transfer function can be evaluated by calculating the numerator and denominator coefficient (for each power of s) for the open-loop transfer function. The numerator coefficients of the open loop will then also become the numerator coefficients for the closed-loop transfer function. The denominator coefficients for the closed-loop transfer function will become the sum of the coefficients of the numerator and denominator coefficients of the open-loop transfer function.

The root-locus analysis approaches the transfer function from the same perspective, wherein the open-loop gain K_O is the lowest-order numerator coefficient of the open-loop dynamics $G(s)$. If K_O is factored out of $G(s)$, we obtain

$$T.F. = \frac{K_O G_n(s)}{G_d(s) + K_O G_n(s)}$$

The denominator contains the characteristic equation. The solution (roots) of the characteristic equation, for all values of K_O, yields the root-locus plot. Since it is a combination of the poles and zeros of the open-loop response, the root-locus plot will determine the stability. Time response also can be extracted from the plot.

The root-locus plot is built upon the computer-derived transfer function, expressed as a ratio of the numerator factors (open-loop zeros) and denominator factors (open-loop poles). The frequency response can be computed from the real and imaginary components of this transfer function. The actual roots for a given gain (K_O) can be obtained through [2]. The frequency-domain, or sinusoidal, transfer function may be written as

$$G(s)\Big|_{s=j\omega} = G_r(\omega) + jG_i(\omega) = G_m(\omega)e^{j\phi(\omega)}$$

where

$G_r(\omega)$ = extracted real component of $G(s)$

$G_i(\omega)$ = extracted imaginary component of $G(s)$

$\phi(\omega)$ = $\arctan\{G_i(\omega)/G_r(\omega)\}$

If the sinusoidal transfer-function program is obtained from the polynomial transfer-function program method, the system's transportation lag cannot be easily handled. If, however, the frequency response is handled on a component-by-component basis, to obtain the magnitude and phase of each as a function of frequency, then the transportation lag (with time lag T_{lag}) can be handled as just another element (with a magnitude of unity and a phase lag of $-360fT_{lag}$). When one evaluates a loop in this fashion, the program generates a series of frequencies at which the loop elements are to be evaluated. This frequency generator has to be set up only once for the program. This style of frequency analysis lends to programming on hand-held programmable calculators.

After the frequencies are evaluated, each loop element and operation (gains, integrators, first- and second-order leads and lags, transportation lag, and loop closure) are evaluated at each of these frequencies. The following method of evaluating the individual elements at each frequency allows each element to be separated into magnitude and phase terms. Since the sinusoidal frequency response is desired, each element is evaluated with s replaced by $j\omega$. Note that "*" refers to the necessary parameters within the subroutine (or array).

Integrator: $1/s$ becomes $1/j\omega$

$$\text{magnitude} = \frac{1}{\omega}, \quad \text{phase} = -90°$$

First Order: $(Ts + 1)^{\pm power} = (j\omega T + 1)^{\pm power}$

$$\text{magnitude} = \{(\omega T)^2 + 1\}^{P_1(*)/2}, \quad \text{phase} = P_1(*)\arctan(\omega T)$$

Second Order: $\left[\left(\frac{s}{\omega_n}\right)^2 + \left(\frac{2\zeta}{\omega_n}\right)s + 1\right]^{\pm power}$

$$= \left[\left(\frac{jw}{\omega_n}\right)^2 + 2\zeta\left(\frac{jw}{\omega_n}\right) + 1\right]^{\pm power}$$

$$\text{magnitude} = \left[\left\{1 - \left(\frac{\omega}{\omega_n}\right)^2\right\}^2 + \left(\frac{2\zeta\omega}{\omega_n}\right)^2\right]^{P_2(*)/2}$$

$$\text{phase} = P_2(*)\arctan\left(\frac{2\zeta\omega/\omega_n}{1 - (\omega/\omega_n)^2}\right)$$

Transportation Lag: $e^{-sT} = e^{-j\omega T}$

magnitude $= \left| e^{-j\omega T} \right|$

$= \left| \cos(\omega T) - j\sin(\omega T) \right| = \sin^2(\omega T) + \cos^2(\omega T) = 1$

phase $= \omega T_{lag}$

Closed Loop [1]:

$$G_c(s) = \frac{G_o(s)}{1 + G_o(s)} = M_c e^{j\phi_c} = X'_o + jY'_o = M'_o e^{j\phi_c}$$

$$G_o(s) = M_o e^{j\phi_o} = X_o + jY_o$$

$$G_c(s) = \frac{M_o e^{j\phi_o}}{1 + X_o + jY_o} = \frac{M_o e^{j\phi_o}}{X'_o + jY_o} = \frac{M_o e^{j\phi_o}}{M'_o e^{j\phi'_o}}$$

$$G_c(s) = M_c e^{j\phi_c}$$

where $X'_o = 1 + X_o$, $M_c = M_o/M'_o$, $\phi_c = \phi_o - \phi'_o$, M = magnitude, ϕ = phase, and o [c] = open [closed] loop, indicating subscript.

Experimental results can be included within a loop if the magnitude and phase are input at the same frequencies established from the frequency generator subroutine. Other methods can be implemented to minimize the number of inputs, some of which average the values in between the requested values. The computer program flowcharts reflect both the polynomial (transfer function in terms of $G_n(s)$ and $G_d(s)$ and the frequency-response (evaluate the dynamic elements at a frequency) scheme.

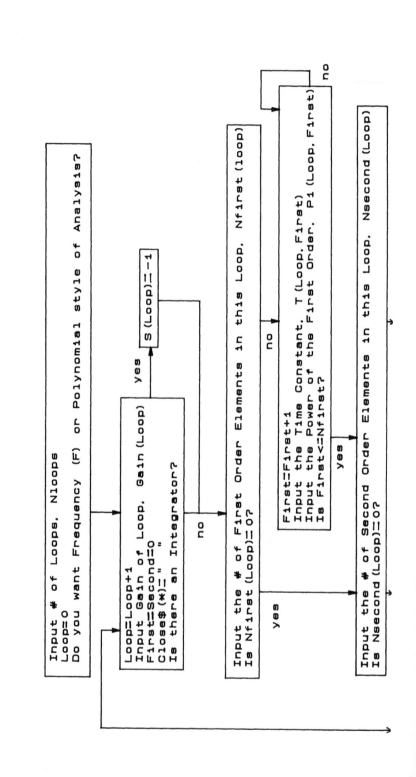

Input # of Loops, Nloops
Loop=0
Do you want Frequency (F) or Polynomial style of Analysis?

Loop=Loop+1
Input Gain of Loop, Gain (Loop)
First=Second=0
Close$(*)= " "
Is there an Integrator?

yes

S (Loop)=-1

no

Input the # of First Order Elements in this Loop, Nfirst (loop)
Is Nfirst (Loop)=0?

yes

no

First=First+1
Input the Time Constant, T (Loop,First)
Input the Power of the First Order, P1 (Loop,First)
Is First<=Nfirst?

yes

no

Input the # of Second Order Elements in this Loop, Nsecond (Loop)
Is Nsecond (Loop)=0?

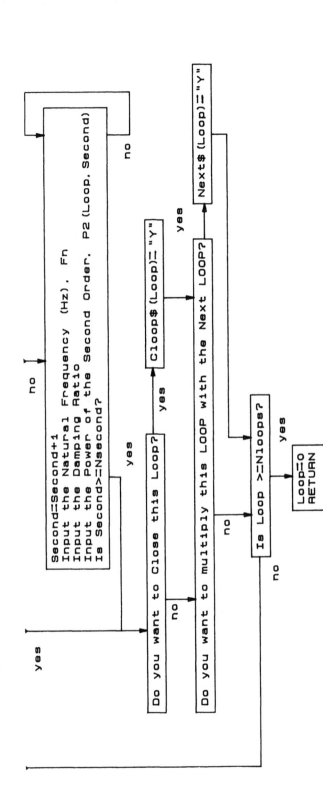

FIGURE A1.4 INPUTS for open- and closed-loop simulation.

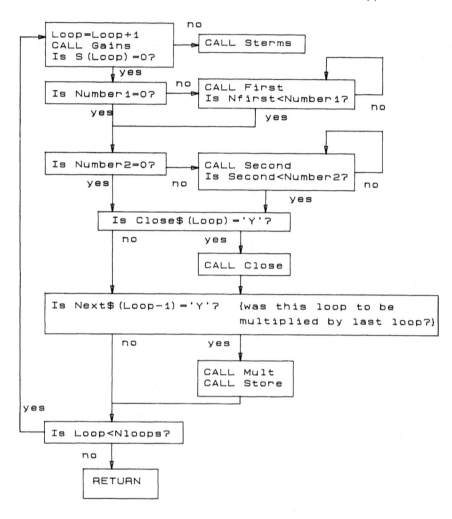

FIGURE A1.5 CALCULATE Main section of loop calculations.

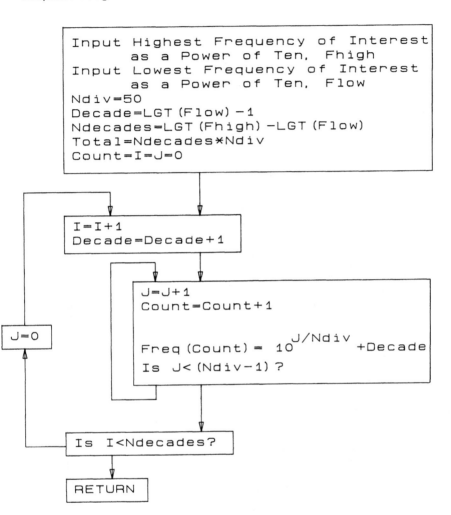

FIGURE A1.6 FREQUENCY GENERATOR Sets up array of frequencies from which loop elements are evaluated.

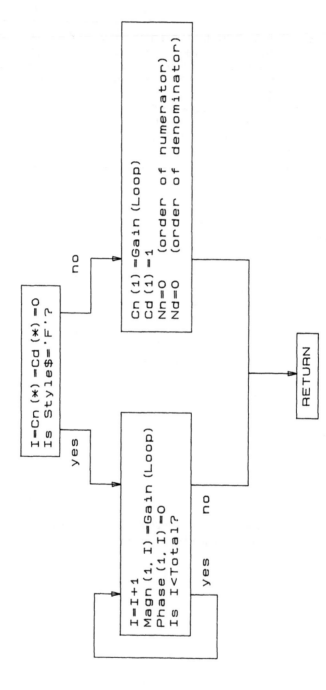

FIGURE A1.7 GAIN Subroutine for evaluating gain of loop.

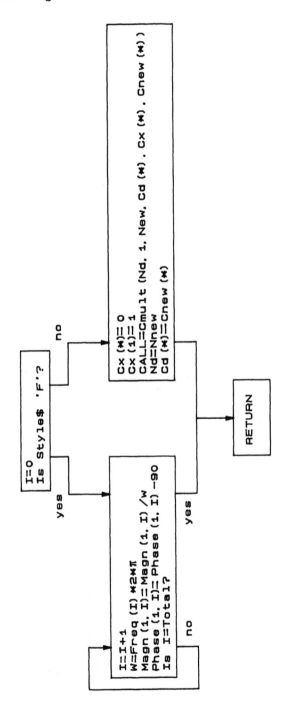

FIGURE A1.8 S-TERMS Integrator subroutine.

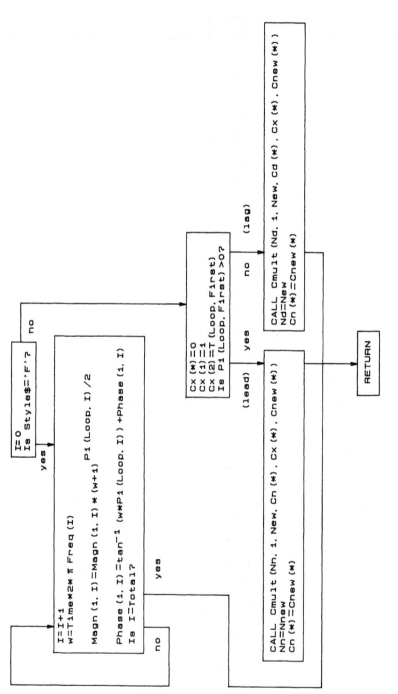

FIGURE A1.9   FIRST   Subroutine for first-order leads and lags.

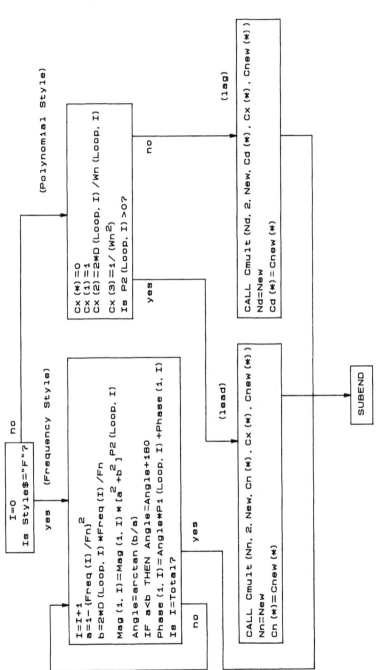

FIGURE A1.10 SECOND Second-order lead and lag subroutine.

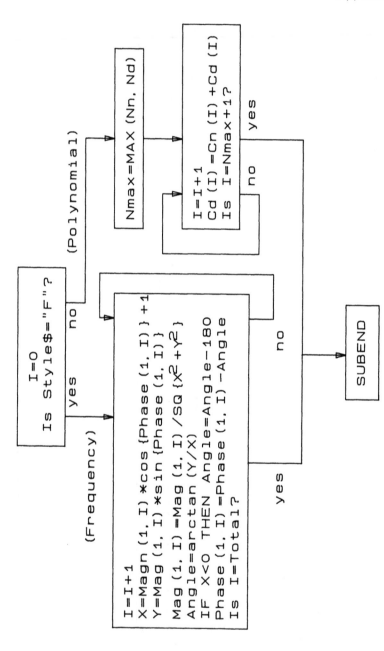

FIGURE A1.11 CLOSE Closed-loop subroutine.

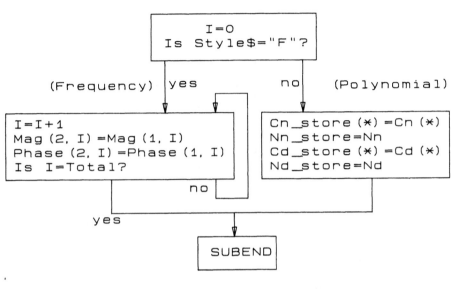

FIGURE A1.12 **STORE** Subroutine which stores loop information when multiplying loops together.

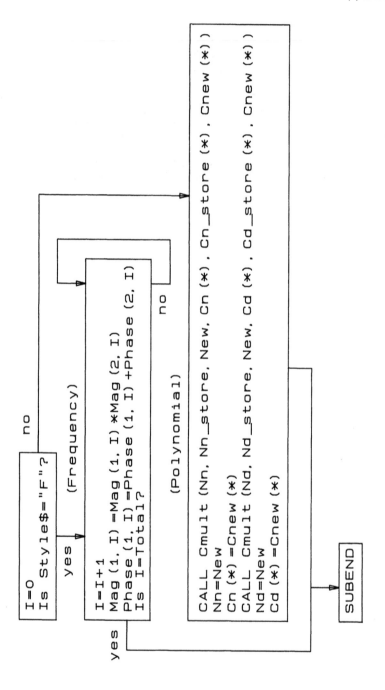

FIGURE A1.13 MULTIPLY Multiplying subroutine for combining loops.

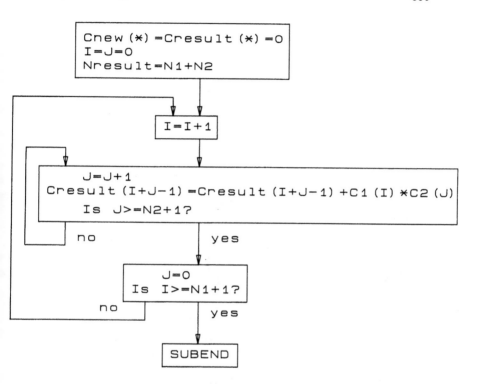

FIGURE A1.14 Cmult Polynomial (coefficient) multiplication.

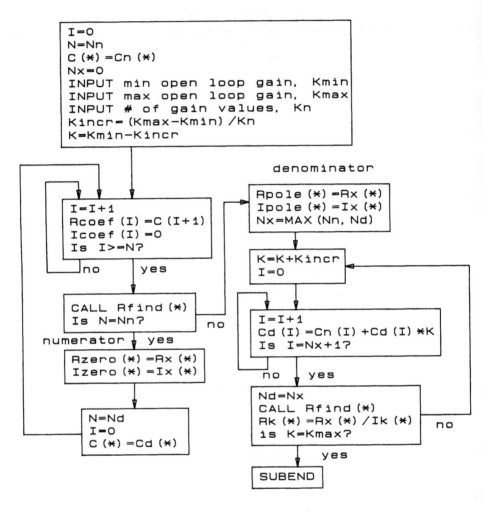

FIGURE A1.15 ROOT Root-locus subroutine.

BIBLIOGRAPHY

1. Wong, L. Gordon. These simple calculator programs replace unwieldy Nichols charts, *EDN*, Nov. 5, 1977, pp. 99--102.

2. Moore, J. B. A convergent algorithm for solving polynomial equations, *Journal of the Association for Computing Machinery*, vol. 14 (1967), 311–315.

Appendix 2
The Microprocessor in Closed-Loop Applications

To experience the benefits of the microprocessor in closed-loop control schemes, the microprocessor must meet the response requirements of the total system in sensing and commanding to the plant. To meet the time demands of closed-loop control, the desired microprocessor is the one which uses both hardware and software schemes to obtain a versatile controller.

The Intel MCS-96 series microcontroller chip fulfills these requirements for closed-loop controls. This chip will be discussed as a controller for obtaining closed-loop response with electrohydraulic plants. Although other microcontrollers fit the requirements for closed-loop applications, this chip is discussed because of the author's familiarity with the product. Programming techniques will be shown in assembly language and the high-level language PL/M in order to implement the compensation procedures of Chapter 5 with total system loops of Chapter 6.

Several levels of programming exist in the microprocessor. The more common high-level languages, such as BASIC, FORTRAN, PASCAL, C, PL/M, and PL/I, are used because of efficiency and/or simplicity. BASIC is the easiest to use, but it is also the slowest and least efficient. Some BASIC programs contain binary support programs to increase the efficiency, and some use compilers instead of interpreters. PASCAL is more highly structured in format but very powerful through its structured set-up. FORTRAN is very popular and almost universally available. Assembly language is one level lower, just one step above the machine language (the computer's method of communication). Assembly language is the most direct organizational means of structuring the machine language into an understandable scheme.

The microprocessor dedicated to control loops is called a microcontroller. Its assembly language is optimized for hardware that performs the necessary functions of the system. The microprocessor, whether fit to control systems or used as a computer, has a characteristic architecture. The microcontroller must be accessed through input and output devices. Timing events, talking to input and output devices, and accessing programs and storage must occur in the proper format, with priorities set by the microprocessor. Within the microprocessor is the crystal (or clock) which performs the correct timing for the protocol and synchronization between the microprocessor and the outside world.

A microprocessor performs some functions at discrete time intervals, whereas other operations may change, depending on program branching and actual computation time. When used as a microcontroller within a closed-loop system, the computation time may have to be minimized in order to meet system demands (where the timing to inputs and outputs is short). In other words, a 20-Hz bandwidth system will demand very short delay times for sampling the analog portion of the system, if the entire system is to keep operating properly. Since there are number-crunching operations involved in the controller algorithms, fast manipulation is essential. Digital-to-analog (D/A) and analog-to-digital (A/D) converters are necessary to convert back and forth between the continuous and discrete systems.

Many designers of electrohydraulic systems are incorporating an interface between analog and digital elements in order to make their components more compatible in terms of cost and ease of implementation. Several valve manufacturers have designed digital input servovalves. Other companies have optimized linear displacement transducers to provide an accurate means of providing digital output from the transducers direct to the microcontroller. Speed and frequency output can be measured and easily converted to digital feedback for the microprocessor. Products such as Intel's microcontroller have pulse-width modulated output signals which can be controlled through software in the microprocessor. The pulse-width output signal acts as a D/A converter for plants which can accept and operate under its signal format.

With this style of output, typical analog electrohydraulic servo-valves (such as those discussed in Chapter 4) can be driven directly as long as the voltage and current demands of the valve are within the operating range of the microcontroller.

The structure of a good microcontroller is such that, in addition to its ease of interfacing, it enables good software control of hardware, along with a fast clock. There are systems which demand too much from the microprocessor in terms of time requirements. A majority of electrohydraulic systems are well suited for a microcontroller, such as the Intel MCS-96 series. Speed enhancements evolve from the faster clock rates, 16-bit (or larger) architecture, register manipulation rather than accumulator usage, hardware high-speed multiply and divide, and high-speed inputs and outputs.

The assembly language builds on the structure of the registers, inputs and outputs, and the timing of the events. The assembly language must accommodate the protocol of the microprocessor and manipulate the data in terms of the binary system rather than the decimal system. The hexadecimal system is utilized to handle the binary bits (0's and 1's) in combinations much like a word is used to represent a mixture of letters.

The bit is the digital signal transmitted throughout the digital control scheme. It occurs either as the binary number 1 (an electrical signal present) or 0 (no electrical signal present). With only two voltage states available (say 5V representing a binary 1), the probability of the states 0 and 1 being disordered is negligibly remote. Other error-detecting schemes further enhance the reliability of the data stream.

It is necessary to perform math functions in binary form. For example, the base 10 value of

$$X_{10} = 234.5$$

is equivalent to

$$(2 \times 10^2) + (3 \times 10^1) + (4 \times 10^0) + (5 \times 10^{-1}) + (6 \times 10^{-2})$$

Generalized to any base, a number is represented as

$$X_B = \cdots + a_3 B^3 + a_2 B^2 + a_1 B^1 + a_0 B^0 + a_{-1} B^{-1} + a_{-2} B^{-2} + \cdots$$

where B is the base of the number. The binary base is 2 with the number representation

$$X_2 = \cdots + a_3 2^3 + a_2 2^2 + a_1 2^1 + a_0 2^0 + a_{-1} 2^{-1} + \cdots$$

This is represented as

$$X = \cdots a_3 a_2 a_1 a_0 \cdot a_{-1}$$

where the dot represents the beginning of negative exponents.

A word on the Intel is 16 bits or 2 bytes. A byte is 8 bits. The microprocessor distinguishes between the upper and lower bits of a word. The 16-bit architecture can describe the word by representing it under a hexadecimal or base 16 system. The hexadecimal is represented by the base 10 numbers plus the letters A through F. For conversion, the same base equation applies.

The following table shows the equivalence between the decimal, binary, and hexadecimal systems. Note, in the lower byte, the individual bit is declared by its power of 2. Therefore, the decimal 8 is 2^3; this puts a 1 at bit 3 of the lower byte (discussion follows). The hexadecimal system makes life easier for handling the binary numbers. Since hexadecimal is a base 16 system, it can represent half of a byte (a nibble) by a single character or number, as shown. Therefore a 16-bit word is represented by a four-position hexadecimal equivalent. The possible values for the lower half of a byte (nibble) are represented below for all three systems:

BINARY	HEX	DECIMAL
3210		
0000	0	0
0001	1	1
0010	2	2
0011	3	3
0100	4	4
0101	5	5
0110	6	6
0111	7	7
1000	8	8
1001	9	9
1010	A	10
1011	B	11
1100	C	12
1101	D	13
1110	E	14
1111	F	15

The hexadecimal number E 2FA is represented as

$$E2FA = E(16^3) + 2(16^2) + F(16^1) + A(16^0)$$

$$= 14(16^3) + 2(16^2) + 15(16^1) + 10(16^0) = 58106$$

The 16-bit word is represented as

16-bit word

upper byte	lower byte
0 1 0 1 1 1 1 0	1 0 0 1 0 1 0 1 = AE95 (hex)
2^{15} 2^{14} 2^{13} 2^{12} 2^{11} 2^{10} 2^9 2^8	2^7 2^6 2^5 2^4 2^3 2^2 2^1 2^0

Note that the highest decimal number which can be represented by the low byte and by the total 16-bit word are, respectively,

byte max = 2^8 - 1 = 255,   word max = 2^{16} - 1 = 65,535

Once the software program is written and edited, it is assembled. The unassembled program is termed the *source* file, whereas the assembled program is termed the *object* file. The assembled version converts mnemonic terms such as LD D,A (load contents of register A into D) into the machine language of 1's and 0's. When in the assembled mode, the program can be executed.

The final configuration of the software is called firmware when the program is placed in ROM (read-only memory), to be used as the control algorithm for the system. While the system is operating, the firmware controller algorithm will use temporary storage registers or RAM (random-access memory) to perform its function within the control loop. A pulse-width modulated (PWM) output signal is desirable for driving many styles of electrohydraulic servovalves.

A microcontroller can have pulse-width output signals driven via hardware or software (within the microcontroller). A PWM signal is a valve-driven signal which is a constant amplitude with varying widths of pulse signals. Therefore, the duration (width) and polarity of the pulse will determine the effective drive signal to the servovalve.

To show how a signal is sampled without use of the A/D or D/A converters, we discuss a program which utilizes a frequency-to-voltage converter to monitor output rotational velocity of a control loop. The timing of both the PWM (output from the controller)

and the frequency-to-voltage converter (input to the controller),
as well as control algorithm, must be implemented within the constraints of the system requirements. The sampling of the output
(speed) must be updated at a minimum rate set by stability and
system response requirements. The remaining control algorithm
output function must allow the sampling to occur at the rate dictated for proper performance.

Before driving the PWM circuit to operate a valve within a
plant, the microcontroller must first compare the command signal
with the output. To obtain a good feel for the microcontroller in
closed-loop operation, we use a PL/M program. Insight into other
microcontrollers, in general, will become apparent through this
high-level language.

The language must provide direct manipulation of the bits controlling the inputs, outputs, and controller settings; it should
also be fairly easy to program. The feedback assumed is a velocity signal, measured by a frequency-to-voltage converter. By
controlling the feedback through the microcontroller, comparing
its signal with the command signal, and driving a valve through
the controller's PWM circuit, many features of the microcontroller
can be employed.

The PL/M program, similar to PASCAL and C, gives a good insight into the hardware and software interplay in a microcontroller. The discussion which follows will show how a microcontroller
loop works, with minimum effort of learning a new language. In
other words, the PL/M program indicates the system operation,
irrespective of the microcontroller utilized. Chapters 5 and 6 use
this background to develop compensation techniques required for
stability control over the plant.

Because a microcontroller is involved in real-time loop development, typical methods of programming must be modified. For example, if a system is to be sampled at a given rate, to satisfy
rules for controllability and stability, background programs (such
as updating information on a display) cannot interfere with the
more time-critical elements. Therefore, interrupts to the main
flow of a program become necessary to fulfill all of the needs of
a microcontroller.

Loop functions (such as sampling the feedback, updating internal loop gains, summing the latest input and feedback, and
outputting the appropriate signal to the plant's electrical interface) must occur at the sampling instants, but background jobs
must only occur after the loop parameters are satisfied. Before
implementing the control structure in PL/M programming language,
we discuss the basic background of PL/M.

Assembly language allows the programmer to directly access
the bits of registers for program storage and for manipulating
inputs, outputs, and general flow structure of a program. PL/
M (which is effectively a superset of an assembly language) also
allows direct bit manipulation in its software, with the advantages of easier and shorter programs (but not necessarily more
efficient) than the assembly software. PL/M is a structured
language based on DO blocks and statements. A DO block is a
set of statements which defines the entire program or a portion
of a program. The main uses of the DO blocks are outlined below:

1. Main Program DO Block

```
MAIN:  DO;
          DECLARE TEST BYTE;
          -----------; /* PROGRAM STATEMENTS */
          -----------; /*      "           "       */
          -----------; /*      "           "       */
       END MAIN;
```

MAIN represents a label for the DO block. The DO block is com-
pleted at the END statement with its reference label. All statements end in ";". Anything within the symbol /*---*/ is a com-
ment to the program lines and will be used in this discussion to
explain the steps. DECLARE is necessary to the program in order to distinguish variable types (word, byte, real, integer,
etc.) and their location. A timer variable is represented as a
WORD, whereas another variable would be used as a BYTE in
order to use less memory (if the program or statements allow
the smaller representation).

2. Subprogram -- PROCEDURE DO Block

```
SUB: PROCEDURE (A);
        DECLARE B BYTE;
        B=10H;                 /* H represents hexadecimal value*/
        A=A+B;
END SUB;
```

This block acts like a subroutine which passes the parameter A
through the procedure and the main program. The variable B
is known only to the procedure DO block. If the passing parameter is neither specified in the first line nor DECLAREd in

the procedure, then the variables used in the procedure are assumed to be from the main program (calling program). The procedure can be slightly modified to enact the function subroutine equivalent.

3. DO Statement

```
IF TEST=1 THEN    /* bit 0 checked -- if =1, then true */
   DO;
      VALUE=100;   /* unlabeled variable is decimal        */
      I=I+1;       /* (10010011B -- B represents binary)   */
   END;
ELSE
   I=I+2;
```

The conditional IF statement and its ELSE expects to be contained in only one statement (where a statement is delimited by the symbol ";"). Placing a DO block at these locations will treat all statements within the DO block as one statement.

4. DO WHILE Statement

```
DO WHILE SETPOINT>FEEDBACK;
   --------;
   --------;
   --------;
   FEEDBACK=FEEDBACK+.001;
   --------;
END;
```

As in the IF statement, when the expression SETPOINT>FEEDBACK is true, bit 0 of the expression is equal to 1 and the block is repeated. Once the variable FEEDBACK is increased to a value which makes the expression false, the low bit of the expression is zero, and the block terminates.

5. DO CASE Statement

```
DO CASE LOOP;
GAIN:  GAIN=80;         /* LOOP=0 */
PID:   DO;              /* LOOP=1 */
          Kp=1200;
          Td=.001;
          Ti=.05;
       END PID;
```

```
PDF:  DO;                    /* LOOP=2 */
         Ki=750;
         Kdl=.003;
         END PDF;
      END;
```

The value of LOOP, if equal to zero, will go to the first state-ment, evaluate it once, and exit the block. Similarly, it would branch to the second (third) statement if the value is equal to 1 (2), and exit.

6. Iterative DO Block

```
DO I=1 TO 10 BY 2;
   A=A+I;
   --------;
   --------;
END;
```

The DO block will be repeated until I reaches the value 10 (with all statements of the block evaluated), wherein I is incremented by 2 for each block repetition.

With this background of DO blocks, the controller algorithm from feedback to valve output command for a proportional loop (no integrator or compensation networks) will be shown. The general flow of information will lead through a main block, and it will be interrupted by an interrupt procedure which updates both the feedback and output values. The following discussion is summarized in Figure A2.1. Portions of these "modules" have evolved through [1].

The main program DECLAREs all variables, including the var-iables associated with the hardware (such as the timers and in-put and output ports). The program will not show the neces-sary DECLARE's of bytes, integers, reals, etc. PROCEDURES are set up immediately after the DECLARE statements. Initial values of variables are then inputs for the main program and subsequent interrupts necessary for loop sampling requirements.

The main program sets up the software timer; this will trigger an interrupt procedure which will address both the output and feedback parameters. The initial part of interrupt will set up the software timer to trigger the next invocation of its interrupt. At each interrupt, the software will send an output command to an electrohydraulic valve.

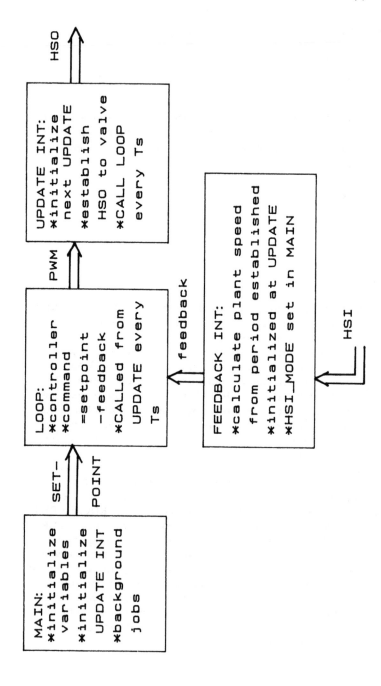

FIGURE A2.1 Digital controller block diagram including interrupt routine implementation.

The feedback values will be updated only at interrupt times which satisfy the sampling time of the loop (which is therefore a multiple of the interrupt timing). The interrupt procedure calls the necessary procedures which fulfill the loop-s requirements at the sampling instants. Once the interrupt procedure is complete, the program execution returns to the point in the main program where the interrupt occurred.

The main program completes background tasks which have lower priority than loop sampling, output assignments, or fail-safe requirements. These may include updating pressure and temperature values for visual inspection on a display board, and data manipulation in establishing these parameters. The program (with hardware registers HSI-MODE, HSI-TIME, etc.) uses HSI (high-speed input), HSO (high-speed output), BITTST (check for the presence of a bit of a byte or word being set to 1), and variations of the commands discussed. Figure A2.1 is a system feedback-style block diagram of the controller with its interrupt routines.

```
CONTROLLER: DO;      /* start of main DO block */

DECLARE (------,------) BYTE;  /* variables declared to match */
                               /* requirements of micro's     */
                               /* hardware                     */
FEEDBACK: PROCEDURE INTERRUPT 2;
   /* this interrupt is established when HSI has made its */
   /* targeted transition as set by HSI-MODE in the main */
   /* program -- it is re-enabled for another period      */
   /* reading of the output (speed) only at the interval  */
   /* of the UPDATE interrupt (otherwise it would be      */
   /* constantly interrupting for high output speeds)     */
DECLARE ------;
   /* status of FIF0, HSI, TIME, etc */
DO WHILE BITTST(.I0S1,7)  /* repeat loop while info avail */
STATUS_HSI=HSI_STATUS    /* save info at interrupt        */
TIME_HSI=HSI_TIME        /* save time at interrupt        */
IF BITTST (.STATUS_HSI,0) THEN  /* check for HSI.0 int */
   DO;
     IF SPEED=0 THEN
       DO;  /* either initial start up or slow speed */
       START=TIME_HSI;
       INPUT_COUNT=1;  /* initial part of the feedback */
       INPUT_OVERFLOW=0;  /* new speed reading to develop */
       END;
```

```
      ELSE
        DO;
        STOP=TIME_HSI;
        CALL BITCLR (.IOCO,0);  /* disable HSI.0
        INPUT_COUNT=0;
        IF INPUT_OVERFLOW<2 THEN  /* very low output speed      */
                                  /* -- see TIMER_OVER_FLOW */
                                  /* procedure                   */
          DO;  /* obtain input speed -- used in LOOP */
              /* which is called from UPDATE inter. */
          LAST_SPEED=SPEED;  /* update */
          SPEED=CONVT*(STOP-START);  /* conversion factor */
          IF COMMAND<0 THEN SPEED=SPEED;
          END;
        ELSE  /* very long period -- out of resolution */
              /* of feedback mechanism                 */
          SPEED=0;
        END;
      END;
END FEEDBACK;

LOOP: PROCEDURE PUBLIC INTERRUPT-CALLABLE;
  /* main control loop items evaluated at sampling periods      */
  /* output for PWM established from feedback, setpoint,         */
  /* and proportional controller gain                           */
  /* CALLED from AT_SAMPLING (DO block) of UPDATE inter. */
  /* sum feedback from HSI with setpoint -- match to PWM        */
  /* circuitry of microcontroller's hardware                    */
  COMMAND=SET_POINT-SPEED;
  /* match command to PWM register protocol which is used */
  /* to drive the HSO (electrohydraulic servovalve) in the */
  /* next invocation of the UPDATE interrupt              */
  IF COMMAND>0 THEN DO;
    POS_ON=COMMAND*MAX_ON;  /* obtain proper drive to valve */
    /* MAX_ON is set to correspond to the maximum rating of    */
    /* the servovalve as well as the buffer amplifier to the   */
    /* servovalve                                              */
    NEG_ON=0;
    END;
  ELSE DO;
    POS_ON=0;
    NEG_ON=COMMAND MAX_ON
  END;
  IF COMMAND=0 THEN (POS_ON,NEG_ON)=0;
  ENABLE;
END LOOP;
```

```
UPDATE: PROCEDURE INTERRUPT 5;
  /* update inputs and outputs at regular intervals   */
  /* inputs will be implemented at a multiple of this  */
  /* interrupt (software) timing -- see AT_SAMPLING */
  UPDATE_START=UPDATE_START+DELTA_UPDATE; /* set up next int */
  /* ------------ perform HSO control ----------- */
  /* note that up to 8 events can be pending at any time */
  HSO_COMMAND=0001$1000B; /* uses software timer0 & inter */
  HSO_TIME=UPDATE_START; /* HSO inter. @ next UPDATE   */
  /* drive valve */
  PWM_BEGIN=UPDATE_START+200;
  HSO_COMMAND=0010$0000B; /* turn on valve in pos dir @HSO.0 */
  HSO_TIME=PWM_BEGIN;        /* just after HSO inter begins      */
  /* keep on until: */
  HSO_COMMAND=0; /* turn valve off at */
  HSO_TIME=PWM_BEGIN+POS_ON
  /* turn valve on neg */
  HSO_COMMAND=0010$0001B; /* turn on valve (neg) @HSO.1 at */
  HSO_TIME=PWM_BEGIN;
  HSO_COMMAND=1; /* turn valve off @HSO.1 at */
  HSO_TIME=PWM_BEGIN+NEG_ON;
  /* set up correpondence clock and sampling */
  /* set up conditional to get into sampling time routine */
  IF (UPDATE-NUM MOD UPDATE-SAMPLE)=0 THEN
    AT_SAMPLING: DO;
      /* enable HSI.0 */
      IOCO=0000$0001B;
      /* clear pending HSI */
      CALL BITCLR(.INT_PENDING,2);
      /* set up interrupt mask for software timer, */
      /* HSI available, and timer overflow */
      INT_MASK=0010$0101B;
      ENABLE;
      CALL LOOP EXTERNAL; /* perform loop requirements */
      CALL LONG_PERIOD;
      /* pulse watchdog timer so that shutdown doesn't occur */
    END AT_SAMPLING;
END UPDATE;

LONG_PERIOD: PROCEDURE PUBLIC INTERRUPT_CALLABLE
  IF INPUT_OVERFLOW>2 THEN DO;
    INPUT_OVERFLOW=0;
    INPUT_COUNT=0;
    SPEED=0;
  END;
END LONG_PERIOD;
```

```
TIMER_OVERFLOW: PROCEDURE INTERRUPT 0; /* >0.13 sec period */
                                       /* of output speed   */
INPUT_OVERFLOW=INPUT_OVERFLOW+1;
END TIMER_OVERFLOW:

/* initialize variables */
/* this is actually the start of the main program */
**
/* set up software timer for initial interrupt */
UPDATE_START=TIMER1+1000;
/* establish HSI_MODE for measuring input transitions */
/* which will be used to calculate output speed        */
/* in the FEEDBACK INTERRUPT ROUTINE                   */
HSI_MODE=0000$0001B;
                /* this acknowledges positive transition of the  */
                /* output (speed) -- the second transition       */
                /* represents one period of the output (speed)   */
                /* (which is calculated at the FEEDBACK inter.)   */
                /* -- it is enabled only at the UPDATE interval   */
                /* in order to get a value only at the UPDATE     */
                /* time -- otherwise the interrupt would be       */
                /* occurring constantly (for high output speeds)  */
                /* using all of the computer's time               */
HSO_COMMAND=0001$1000B; /* timer1 & set inter software timer0 */
HSO_TIME=UPDATE_START; /* time of HSO.0 inter with command */
INT_MAST=0010$0000B; /* HSO, UPDATE, and software timer inter */
ENABLE;
TEMPERATURE=(.TEMP);
PRESSURE=(.PR);
BACKGROUND: DO WHILE 1; /* the expression 1 will always   */
                        /* be true; therefore, the block  */
                        /* will be looped continuously     */
    IF UPDATE_NUM MOD 64=0THEN
    CALL DISPLAY(TEMPERATURE, PRESSURE) EXTERNAL;
    /* perform other background jobs   */
    /* as long as block not interrupted */
END BACKGROUND;


END CONTROLLER;
```

These PL/M programs are used in Chapters 5 and 6. Chapter 5 develops a digital filter using PL/M with an equivalent description in assembly language. Chapter 6 expands on both this appendix and Chapter 5 to provide a closed-loop, digital-compensated, control system.

BIBLIOGRAPHY

1. Burns, Dennis. Personal communication, Sundstrand-Sauer, Minneapolis, Minnesota.

Appendix 3
s-, t-, and Z-Domain Properties and Equivalence

g(t)	G(s)	G(z)
unit step	$\dfrac{1}{s}$	$\dfrac{z}{z-1}$
unit impulse	1	1
exponential $e^{-\acute{a}t}$	$\dfrac{1}{s+\acute{a}}$	$\dfrac{z}{z-e^{-\acute{a}T}}$
$\sin(\omega t)$	$\dfrac{\omega}{s^2+\omega^2}$	$\dfrac{z\,\sin(\omega t)}{z^2-2z\,\cos(\omega T)+1}$
$\cos(\omega t)$	$\dfrac{s}{s^2+\omega^2}$	$\dfrac{z(z-\cos(\omega T))}{z^2-2z\,\cos(\omega T)+1}$
$e^{-\acute{a}t}\sin(\omega t)$	$\dfrac{\omega}{(s+\acute{a})^2+\omega^2}$	$\dfrac{z\,e^{-\acute{a}T}\sin(\omega T)}{z^2-2z\,e^{-\acute{a}T}\cos(\omega T)+e^{-2\acute{a}T}}$
$e^{-\acute{a}t}\cos(\omega t)$	$\dfrac{s+\acute{a}}{(s+\acute{a})^2+\omega^2}$	$\dfrac{z^2-z\,e^{-\acute{a}T}\cos(\omega T)}{z^2-2z\,e^{-\acute{a}T}\cos(\omega T)+e^{-2\acute{a}T}}$

g(t)	G(s)
$\dfrac{\omega n}{\sqrt{[1-\varsigma^2]}}\,e^{-\varsigma\omega nt}\sin\omega_n\sqrt{[1-\varsigma^2]}t)$	$\dfrac{1}{(s/\omega_n)^2+(2\varsigma/\omega_n)s+1}$
$\pounds[Ag(t)]$	$A\,G(s)$
$\pounds[e^{-\acute{a}t}g(t)]$	$G(s+\acute{a})$
$\pounds[g1(t)+g2(t)]$	$G1(s)+G2(s)$
$\pounds[g(t-\acute{a})1(t-\acute{a})]$	$e^{-\acute{a}s}G(s)$
$\pounds[\dfrac{d\,g(t)}{dt}]$	$sG(s)-g(0)$
$\pounds[\dfrac{d^2g(t)}{dt^2}]$	$s^2G(s)-s\,g(0)-g(0)$
$\pounds[\int g(t)dt]$	$\dfrac{G(s)}{s}+\dfrac{\int g(t)dt}{s}\Big|_{t=0}$

FIGURE A3.1   s-, t-, and Z-domain properties and equivalence.

Appendix 4

Block-Diagram Reduction

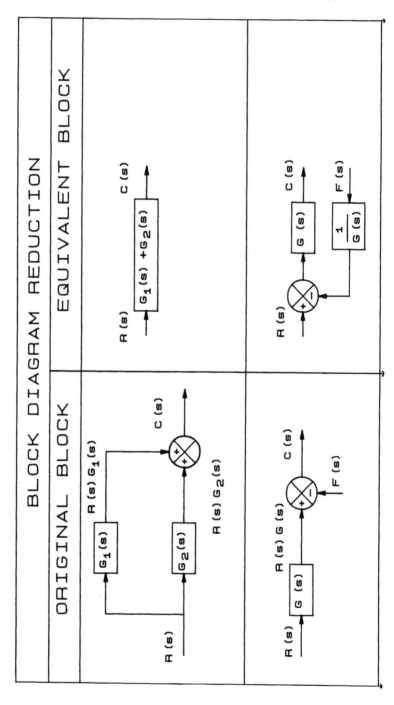

BLOCK DIAGRAM REDUCTION

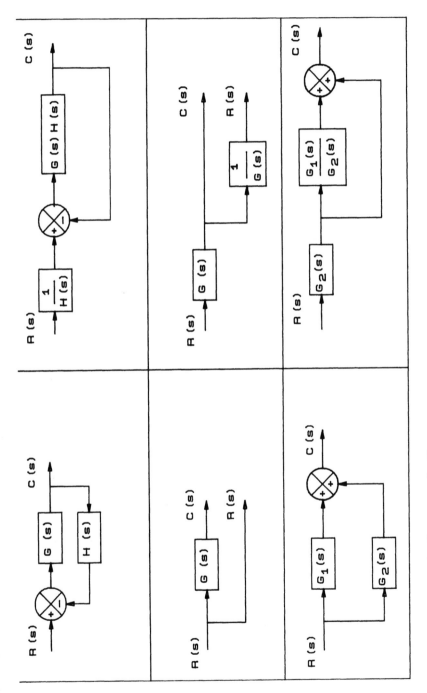

FIGURE A4.1 Block-diagram reduction.

Index